PLUMBING

This book explains the fundamental principles of plumbing, and discusses the basic manual skills it requires and their application on the job. The text is illustrated throughout. Whilst its purpose is primarily to instruct the beginner, the book also aims to fill the gap between theory and practice for the apprentice.

This revised edition incorporates S.I. units but gives their Imperial equivalents in brackets so as to facilitate comparison with existing materials and installations.

D1362848

TEACH YOURSELF BOOKS

PLUMBING

John Hosking Innes
F.I.O.B., M.I.P.H.E.

TEACH YOURSELF BOOKS
Hodder and Stoughton

To Y.D.D.

First Printed	1952
Second edition	1971
Second impression	1973
Third impression	1974
Fourth impression	1976

Copyright © 1971 Edition
Hodder and Stoughton Limited

ISBN 0 340 05689 4

Printed and bound in England
for Teach Yourself Books,
Hodder and Stoughton, London,
by Hazell Watson & Viney Ltd, Aylesbury

GENERAL EDITOR'S FOREWORD

THE AIM of the building section of the Teach Yourself Series is to assist those who are desirous of acquiring information concerning building methods and practice.

It is not intended that these books will take the place of textbooks or recognised courses of study at Technical Colleges, but they should appeal to all students of building because each volume has been written by a specialist in his own particular subject.

The series covers almost every branch of the building crafts and allied professional practice.

In placing before the public this comprehensive work on Building, no apology is necessary for continuing to describe and illustrate traditional methods of building construction, because it is of vital importance that the layman who desires to become acquainted with building technique should be instructed in the basic principles of building.

Although many changes have taken place in the design and layout in post-war systems of plumbing and house sanitation, the basic principles continue to be incorporated and, indeed, these principles are essential in good design.

The author of this book has explained very fully the various items of plumbers' work which are so often enveloped in mystery.

This explanation will assist the reader in obtaining a knowledge of the fundamental principles.

The text has been written in a concise and interesting manner, and the line drawings are clear and self-explanatory, thus enabling the reader who has the practical faculty of reading drawings and the capacity to retain, to gather a wealth of information on this very important section of building technique.

SPECIAL ACKNOWLEDGMENTS

The Publishers are indebted to the following firms for kind permission to use their products in certain illustrations in this book:

Wm. Barton & Sons: "Kongrip" compression fitting. Donald Brown Ltd.: "Brownall" compression fitting. Fyffe & Company Ltd.: Instantor compression fitting. Samuel Gratrix Ltd.: "Buttrix" compression fitting. Hull Steel Radiators Ltd.: double panel radiator. Ideal Boilers and Radiators Ltd.: capillary fitting. Ingersoll-Rand Pumps Ltd.: "Thermopak" fixed head circulator.

The author wishes to offer his sincere thanks to Dr. John Bock for his help on metrication and the section on small-bore heating.

CONTENTS

INTRODUCTION

WITH SOME guidance in simple principles and procedure and an opportunity for practice, there is much in plumbing that you can teach yourself. This book aims at giving such guidance.

It is not possible in a book of this size, nor is it desirable for the beginner, to attempt to deal fully with the technology of plumbing; rather has the emphasis been laid on the operational aspects of the craft.

The plumbing-craft skill needs development along two main lines: by way of the basic manual skills and their application on the job, and by an understanding of simple scientific principles.

To meet this need, so far as possible, actual plumbing operations have been dealt with, described, and illustrated; whilst at the same time the important fundamental principles have been explained.

In recent years there has been a marked departure from the use of the traditional lead materials on new plumbing work and an increased use of copper and other hard metals. In view of this and of the importance and amount of repair work to existing lead installations, space has been devoted to a consideration of both kinds of work.

For apprentices this book might profitably fill a gap between theory and practice. It is not intended as a substitute for a systematic study of the craft, but gives more than usual space to the important aspect of fixing—of how to go about the job.

Students are advised to clear up any obscure points as they arise by consulting text-books, works of reference, the publications of the trade organisations, and the information sheets and documents available from the development associations for lead, copper, tin, zinc, etc.

S.I. units have been used where measurements or forces, etc. are required, but the old Imperial figures have been given in brackets so as to make possible comparison with existing materials and to facilitate repairs or alterations to existing installations. It would seem that, for the foreseeable future, plumbers and other services engineers will require both systems of units.

We should like to express thanks to the Coal Utilisation

Council for the illustration on the small-pipe heating system, and the arrangement of the boiler and control system; and to Fry's Metal Foundries Ltd. for the excellent photographs on soldering and joint-wiping; and to the Lead Industries Development Council for the photographs on sheet lead.

CHAPTER I

THE PRINCIPAL METALS USED IN PLUMBING

HISTORICALLY, the word plumber was used to describe the craftsman whose main function was the manipulation of lead in its various forms. The name is derived from the Latin *plumbari*, a worker in lead, *plumbum*.

Although the plumber is almost the only craftsman today concerned with the working of lead, the discovery of the usefulness of other materials, their ever-increasing availability, and the increase in the cost and importation of lead have tremendously widened the scope and versatility of the craft. Today the best plumbers have a knowledge and skill deserving of greater social and tangible recognition.

Real efficiency in the craft requires a knowledge of physics and chemistry, metallurgy, hydrostatics and hydraulics, and mechanics. This scientific knowledge is best acquired through experience of real problems in actual practice, greatly aided by good teaching in a technical college.

Lead

This soft and malleable metal has been used for some thousands of years. Its use was known in Ancient Egypt, in Greece, and more widely in the Roman Empire, whose colonising Legions were responsible for disseminating so many useful arts. It is to them we owe its early uses in Britain.

The chief source of commercial lead is the ore Galena, which is the sulphide of lead (PbS). This ore is found in almost every part of the earth's surface, with heavy deposits in Australia, Germany, and Spain. Lead also occurs in the form of Cerusite ($PbCO3$) in many places. This is a carbonate of lead.

Galena, which is grey in colour, contains, besides metallic lead, impurities in varying quantities, such as sulphur, silver, antimony, zinc, and copper. These impurities are generally harmful to the finished manufactured lead products, and they are removed after the smelting of the ore. This is done by adding a series of fluxes, each of which combines with one of the impurities, is precipitated at the surface, and can then be removed by skimming. In recent years more careful control of these impurities has led to the production of a greatly superior standard in the quality of sheet lead, which is claimed to have a purity of 99·99%. Traces of zinc, copper, arsenic, or antimony harden lead considerably, and create difficulty for the plumber who has to work it.

Lead melts at about 327° C (621° F), depending on the degree of purity, and alloys readily with many other metals. The most important alloys to the plumber are the solders produced by a mixture of lead and tin; these will be discussed in some detail later under Soldering and Joint Wiping.

Lead is highly resistant to the atmosphere and most acids; it can be said to be almost indestructible, and examples of well-preserved, exposed leadwork 1,000 years old can be found. It should be noted however that tannic acid in oak timber rapidly destroys lead, the reaction producing a white powder which is a lead tannate. As a consequence of this, care should be taken that sheet lead is not laid on oak unless an insulating layer of bituminous felt or building paper is laid between.

Lead is, in fact, susceptible to initial chemical action by the atmosphere. Oxygen and carbonic acid in the atmosphere react with the surface lead, a chemical change takes place, and the lead is coated with the grey film of lead oxide and carbonate, seen on all lead not freshly cut or cast. It is this grey film which prevents any further chemical destruction of the lead by the atmosphere.

Milled sheet lead as used today, replacing the ancient form of cast sheet, is a very superior product; it is uniform in thickness and of maximum workability. It is produced by rolling a plate of lead between two sets of rollers until it reaches the required thickness.

Lead pipe is made in a hydraulic press. When lead is subjected to a pressure of 495 MN/m² (32t/in²), it will melt at ordinary temperatures. In the lead-pipe press advantage is taken of this fact and the lead so liquefied is forced through the annular space provided between a die and a mandrel. In practice, lead is allowed to just set in the press, and pressing commences at approximately 330° C (600° F). On emergence from the die, it immediately solidifies because of the reduced pressure. Any diameter of bore or thickness of wall can be had by merely changing the die and the mandrel. In recent years it has been found advantageous to harden lead for use as water pipes, owing to the tendency of ordinary soft lead to sag between fastenings. For this end a small quantity of copper or tellurium is added.

Copper

This lustrous red-brown metal was also known and used in ancient times. It has played an important part in plumbing for many years now, but with changes in our economic situation it seems as though copper might supersede lead as the basic material of plumbing.

Copper is tough, it can be rolled into thin, durable sheets, and can be drawn into thin-walled tubes—capable of withstanding high pressures—because of its ductility.

Copper has a great advantage over lead in that it does not require such substantial support, by virtue of the lightness of the sheets used in roof-work and the lightness and rigidity of copper tubes. Copper is found all over the world, sometimes in the metallic state, but more usually as one of the ores: Copper Pyrites, Ruby Ore, or Malachite. The chief impurities are iron and sulphur. These must be extracted and a high degree of purity is achieved; modern sheet and tube are of excellent quality and reliability.

Copper, like lead, will withstand weathering and has an almost limitless life. The atmosphere produces a protective green coating or "patina," which is characteristic of weathered copper, and a quality which is utilised for æsthetic and architectural ends.

Copper melts at 1,083° C (1,981° F).

Zinc

Zinc is used in some parts of the country for roof-work in situations where lead or copper would otherwise be used. It is also used in the galvanising of iron tanks and cisterns. This metal has a low melting-point—418° C (784° F).

This blue-grey metal is fairly hard, but can be worked without great difficulty; it can be soldered with ordinary fine solder, using hydrochloric acid (unkilled spirits of salts) as a flux.

Zinc is unsuitable for weathering buildings in an industrial atmosphere because of the detrimental effect of sulphuric acid, one of the products of incomplete combustion of fuel.

The chief ores from which zinc is smelted are: Zinc Blende (ZnS), the sulphide of zinc; Zinc Spar ($ZnCO_3$), the carbonate; and Red Zinc Ore (ZnO), the oxide.

Brass

Brass is an alloy of copper and zinc. Yellow brass, which is commonly used in taps, valves, and other fittings in plumbing, consists of about 70% copper and 30% zinc. For better-quality fittings to withstand high pressures a greater proportion of copper is used.

The Plumber's Tools

By the terms of a National Joint Agreement between the employing and operative plumbers, each craftsman is expected to provide the following kit or "bass" of tools. A weekly tool allowance is paid.

Adjustable spanner 300 mm (12 in)
Allenkeys, one set
Bent pin or bolt
Brace and twist drill up to 10 mm ($\frac{3}{8}$ in)
Blow lamp and gas torch (excepting the replacement of spare parts)
Bobbins, all sizes to 32 mm ($1\frac{1}{4}$ in)
Bossing stick
Boxwood dressers (large or small)
Boxwood setting in stick
Boxwood bending dresser
Boxwood mallets (large or small)
Bradawl
Caulking tools for ordinary work
Chasewedge
Compasses
Cutting pliers
Expanding bit up to 50 mm (2 in)
Footprints, 150 and 230 mm (6 and 9 in)
Fixing points (or clamps)
Flat chisel for wood
Gimlet for lead pipe
Gimlet for wood screws
Gouge for wood
Glass cutter and putty knife if glazing is normally done by plumbers in the district

Hacking knife
Hacksaw frame
Hammers, small and large, 1 kg (2 lb) maximum
Handsaw
Handbrush
Knife, large pocket
Lavatory union key
Pliers, two holes
Plumb bob and chalk line
Radiator union key
Rasp
Rule
Screwdrivers (large and small)
Shave hooks
Small brick drill
Snips
Spirit level
Springs for bending 12·5 mm ($\frac{1}{2}$ in) and 19 mm ($\frac{3}{4}$ in) light gauge copper pipes
Steel chisels for brickwork up to 508 mm (20 in) long
Stillson or other pipe wrench up to 300 mm (12 in)
"Surform" rasp 230 mm (9 in)
Tan pins up to 50 mm (2 in)
Tank cutter
Tool bag and box
Towel, small
Tube cutters or junior hacksaw
Whitworth spanners 6·5 mm ($\frac{1}{4}$ in) to 12·7 mm ($\frac{1}{2}$ in)
Wiping cloths

The employer shall provide the operative plumber with bending springs suitable for use with thinner wall copper tube of up to 19 mm ($\frac{3}{4}$ in). The maintenance and replacement of these springs shall be the responsibility of the operative plumber.

Storage of Tools: Where practicable and reasonable on site, job or in a shop the employer shall provide an adequate lock-up or lock-up boxes, where tools can be left at the owner's risk, except that the employer shall accept liability for any losses caused by fire.

CHAPTER II

BOSSING OR WORKING
SHEET LEAD

LEAD is most useful for flashing and weathering roofs because, in addition to its impermeable and lasting qualities, it is extremely malleable. A malleable metal is one that can be pressed or hammered out of form without tendency to return to it or to fracture. It can be said to be adaptable. A sheet of lead can be manipulated to almost any shape, to fit into any corner, or around any projection through a roof or into a gutter.

This manipulation of sheet lead is known as bossing and involves a considerable degree of skill. In order to be really successful in the development of this skill, two important factors need to be considered. Firstly it is necessary to understand exactly what is required having regard to the properties of lead. Secondly a high degree of mental and muscular co-ordination has to be achieved—what is commonly called the skill of the hands.

Why is lead malleable? What happens when we change its shape? It might be useful to look at the structure of matter in general so that we can understand the nature of lead in particular.

All matter, whether it be solid, liquid, or gas, consists of infinitesimal particles called atoms or molecules—depending on whether it is an element or a compound of two or more elements. As commercial lead is probably never perfectly pure, we will talk of its molecules. In a gas or a liquid fluidity is due to the freedom of molecules to move about. In heavy oil or molten lead this freedom is slightly restricted by a force acting between the molecules, which we call cohesion. In solids this force is much greater, and tends to hold the molecules at a fixed distance from each other, producing rigidity.

It will be obvious that the application of heat or an external force can alter the shape of a solid; the ease with which this can be done will depend on the malleability of the material. Some materials are said to be brittle at normal temperatures; when force is applied fractures or powdering result, unless heat is used first to make them malleable by annealing.

Lead is malleable at normal temperature, and by the careful direction of force its molecules can be moved into any desired

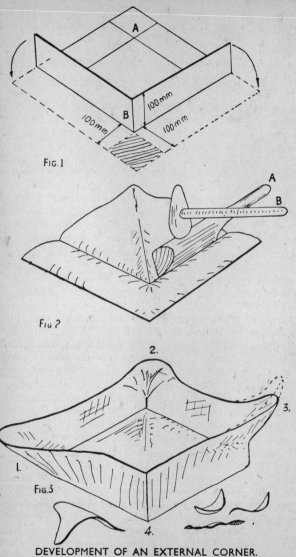

Fig. 1

100 mm

100 mm

100 mm

A

B

Fig 2

A

B

Fig. 3

1.

2.

3.

4.

DEVELOPMENT OF AN EXTERNAL CORNER.

position. As with, say, "Plasticine," the molecules of lead can be moved over, under, and around each other.

Consider why this is necessary. Let us take an example (Fig. 1). Here a flat sheet of lead is being converted into a tray or box with 100 mm (4 in) sides. The corner (A) shows the setting-out before bossing; at (B) we have a finished vertical corner. If the corner (B) is now cut through its depth and the sides are laid flat, it will be seen that the crossed area has been removed. The basic principle in sheet-lead bossing is to be found in the solution of this problem.

The External Corner

The setting-out of the sheet is shown in Fig. 1, the setting-in is illustrated in Fig. 2. The tools needed are a flat dresser (A) and a mallet (B), in the case of thick lead. These can be bought or made and should be of hardwood, such as Lignum Vitæ, Boxwood, or Beech; the mallet handle is most usually of cane. Flat dressers, sometimes called beaters, are right- and left-handed, and with one angle more acute than the other.

The importance of careful preparation cannot be over-stressed —accurate setting-out and firm setting-in—ensure goodness of fit on the job. Measurements carefully taken should be adhered to.

The method of setting-in is shown. This should be carried out on a piece of board to avoid damage and thinning. The angles should be firmly creased and the corner well established at its base by raising the inside of the corner from underneath. Unless care is taken with this operation, the squareness of the corner will be lost, and the worker will be tempted to try to recover later by the use of a sharp-edged tool on the inside. This can only lead to disaster or at least a serious weakening of the lead by stretching. Damage is often done in this connection by the use of a setting-in stick. It is inadvisable that any learner should use or even possess such a tool.

The preparatory operations having been completed, the bossing of what is called an EXTERNAL corner can proceed. The method is shown in four steps in Fig. 3. The additional tools needed are a bossing stick and a pair of snips or hand-shears (Fig. 55). The bossing stick is of hardwood and is illustrated in Fig. 5.

Step 1 is concerned with the raising of the sides to a vertical position and the establishment of the base of the corner. To do this, a number of thickening strokes are used. These are applied

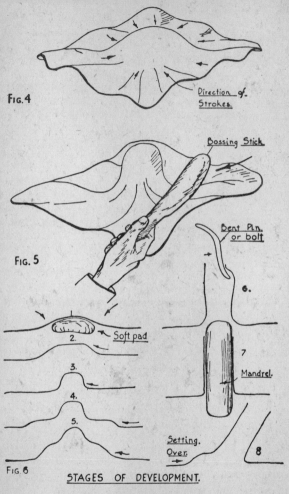

Fig. 4

Direction of Strokes.

Bossing Stick

Fig. 5

Bent Pin or bolt

6.

2.   Soft pad

3.

4.

5.

Fig. 6

7   Mandrel.

Setting Over.

8

STAGES OF DEVELOPMENT.

BOSSING A LEAD-SLATE FLASHING.

in a downward and inward direction until the corner is about an inch high, as seen at 2, Fig. 3.

Step 2.—In order to remove the surplus lead (Fig. 1), a number of stretching or thinning strokes will be needed and, in order that the lead shall be of uniform thickness, they will have to be thoughtfully distributed. From this stage on, the head of the mallet is held along the length of the inside of the corner and is used as an anvil upon which the lead can be stretched, by strokes of the bossing stick, working on the outside. A number of strokes are used in raising the corner, and these will alternate as required. Some will be directed inwards from right and left to keep the corner square, and some will carry the surplus lead upward in waves. Periodically as the bossing proceeds surplus lead should be clipped off. Any lead in excess of the 100 mm of height is an encumbrance and makes the work more difficult. It is interesting to weigh the clippings to see if the surplus area has, in fact, been removed. Great care should be taken throughout that no creases are allowed to develop. Remember that a crease is a potential crack.

Step 3 is a continuation of Step 2, the cutting away of the surplus is shown.

Step 4 shows the final squaring up and trimming for height. From time to time during the working of the corner the sides will tend to sag outwards, and should be straightened up.

When all the corners have been worked and the sides have been dressed, the box should be checked for size—before any attempt is made to square up the angles. The upward bulges in the bottom of the box must not be dressed out until the lead is in position on the job. They will then serve to expand the lead so that it fits more snugly in its permanent position.

The Internal Corner

An internal corner in leadwork usually occurs at a projection into or through a lead-covered flat roof, or where there is a break in the line of a parapet wall gutter.

The bossing problem in this case is much more complicated than that discussed above. An internal corner generally has to be worked in conjunction with one and sometimes two external ones, this is called a break. And the problem is not one of removing surplus lead, but either that of making good a shortage or moving lead from one place to another. In view of what has been

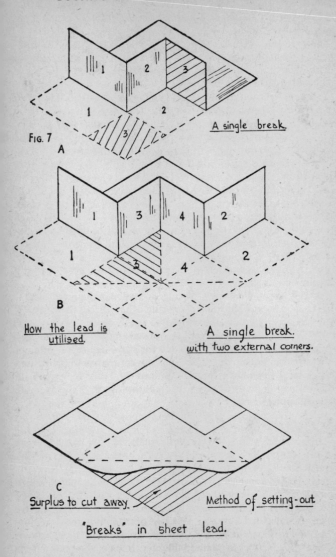

Fig. 7

A

A single break.

B

How the lead is utilised.

A single break.
with two external corners.

C

Surplus to cut away.   _Method of setting-out_

"Breaks" in sheet lead.

said about the nature of lead and the principles of bossing, given time, patience, and energy, lead can be thinned and thickened and driven in waves almost any distance. A good example of this process is illustrated in Figs. 4, 5, and 6.

Depicted here we have a lead-slate, sometimes known colloquially as a "tall hat." Here the lead is driven in concentric waves, from the outside edges of the sheet to the centre. The sketches show how a surplus of lead is brought to the centre and raised in the form of a hollow cylinder or pipe. A subsequent step shows how the upstanding part of the flashing is inclined to the angle at which the pipe for which it is intended penetrates the roof.

A simple calculation will show that a flashing for a 100 mm (4 in) pipe with an upstand of 150 mm (6 in) will contain about 450 cm² of lead in the upstand. This apparently rises from a circle with an area of about 80 cm²; it will be obvious that the extra 370 cm² must be driven in from all sides.

In the case of the internal corner, most trouble arises from the temptation to make progress by using too many stretching strokes. If the corner is to be satisfactory and the lead of uniform thickness, it should almost grow out of the work done around it, and stretching strokes must be limited to these needed to distribute the lead brought in from the sides, or worked down into the corner.

All the problems involved in the working or bossing of sheet lead are to be found in the three jobs discussed above, but only practice and experience can give any reality to the principles and methods described. Much of the skill and the knowledge of the expert craftsman lies in his highly developed muscular and mental co-ordination. It depends on the quality of his "feel" for the lead; he can estimate the thickness of the lead at any point by the sound produced by a stroke of the bossing stick, or by the impression of resistance received by his muscular sense.

The problem involved in the bossing of a "single break" is demonstrated in Fig. 7 (A). Here it will be seen that there is sufficient lead in the flat sheet to produce the upstand 1, 2, 3. But whereas pieces 1 and 2 merely need turning upwards, piece 3 will have to be moved.

Similarly, Fig. 7 (B) illustrates this problem—rendered even more difficult to solve—in the case of a single break with two external corners. Here pieces 1 and 2 can be turned up, but pieces 3 and 4 must be transformed from triangular pieces. Fig. 7 (C)

External Corner using the Bossing Stick.

External Corner Bossing with the Mallet.

PLATE 1

23

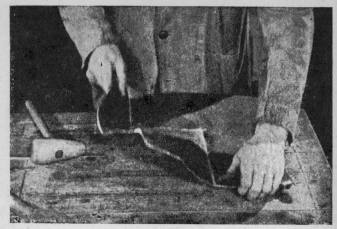

Internal Corner or Break. First Stages.

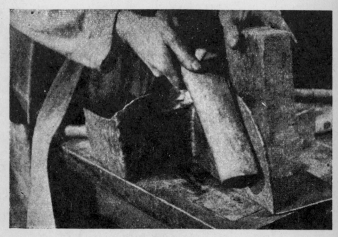

Internal Corner or Break. Final Stages and Straightening up.

PLATE 2

indicates the portion of the sheet which can be cut away as surplus at the beginning. The method of setting-out is also shown. In practice, of course, such a break would occur in a gutter or in some such position, and the lead would extend on each side. Because of this, there would be additional lead available for working towards the angles. In the main, the process is one of driving waves of lead; from the pieces 1 and 2, Fig. 7 (B), and then upwards and back into the angle. It is soon evident when the lead is being stretched. When this is seen, more lead should be worked round from the surplus at the two external angles, 1/3 and 4/2. It is important that the beginner should exercise patience. Most jobs are spoiled by hurrying and the use of too many stretching strokes.

CHAPTER III

LEAD DAMP-PROOF COURSES IN FOUNDATIONS, AT OPENINGS, IN PARAPET WALLS, AND IN CHIMNEYS

A DAMP-PROOF course is a layer of impervious material inserted in a wall to prevent the movement of moisture up, down, or through its thickness.

The ill-effects of rising dampness from foundations are well known: damage to interior decoration, discoloration of facing-bricks, efflorescence, frost damage, and dry-rot are familiar problems. Similarly, dampness may penetrate downwards in a parapet wall or inwards from the face where a lintel or an arch over an opening supports both skins of a cavity wall, thus crossing the cavity. The cavity itself is a vertical barrier to the penetration of dampness, and care must always be taken that it is kept clear of any obstruction which will allow the passage of moisture from the outside weathering skin to the inside of the wall.

Where a house chimney has a large cross-sectional area, it is often recommended that a damp-proof course should be inserted as near as possible to the line at which the chimney emerges.

Damp-proof Courses at Wall Footings

Fig. 8 is an isometric cross-section of a cavity-wall foundation. It will be seen that the D.P.C. is laid at a level between that of the ground-floor and the level of the earth outside. This positioning of the D.P.C. is of the utmost importance and the subject of by-law regulation. Most by-laws require that the D.P.C. shall be below floor level and at least 150 mm (6 in) above the level of the ground outside.

A great variety of materials can be used for the D.P.C., but the plumber is concerned directly with the proper laying of the metallic-sheet D.P.C. These can be of lead or of copper.

Lead D.P.C.

Sheet lead for this purpose varies in thickness according to the quality of the job and the pocket of the customer. It is available

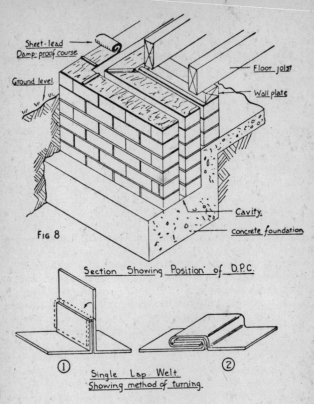

Sheet-lead Damp-proof course

Ground level

Floor joist

Wall plate

Cavity.

concrete foundation

FIG 8

Section Showing Position of D.P.C.

① ②

Single Lap Welt.
Showing method of turning.

FIG. 9

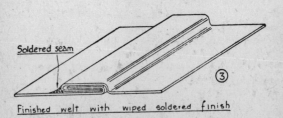

Soldered seam

③

Finished welt with wiped soldered finish

in thickness which varies with its weight per square metre. The Lead Development Association has published concise information about lead sheets in Metric units. An extract of this information is given below in table form for lead sheets in common use.

WEIGHT, THICKNESS AND COLOUR MARKINGS OF LEAD SHEETS

BS Code Number	Weight		Thickness		Suggested colour markings
	kg/m²	lb/sq ft	mm	in	
3	14·18	2·91	1·25	3/64+	Green
4	20·41	4·19	1·80	5/64−	Blue
5	25·40	5·21	2·24	3/32−	Red
6	28·36	5·82	2·50	3/32+	Black
7	34·73	7·33	3·15	1/8−	White
8	40·26	8·26	3·55	9/64−	Orange

The lead commonly used varies between BS3 and BS5, and, as even very thin sheet lead is infinitely more durable than the commonly used bituminous felt, BS3 lead can be considered satisfactory. It will be obvious, however, that BS4 or BS5 lead will endure even longer and withstand more easily any ill-usage to which it might be subjected whilst building proceeds.

In the 275 mm (11 in) cavity wall construction it is usual to lay the D.P.C. in widths of 115 mm (4½ in) on each single brick wall. It is convenient, therefore, to have the lead prepared in rolls of this width. Usually lead sheets are 2·4 m (8 ft) wide and are either of 3 m (10 ft), 6 m (20 ft) or 9 m (30 ft) length. Thus 21 rolls of that width can be cut from a 2·4 m wide roll; where large quantities of such rolls are needed, it should be remembered that manufacturers are willing to supply rolls of any width and of the above lengths. As the joining of separate pieces takes time and adds to labour costs, this economy needs no explaining.

Procedure for Laying and Jointing

In order to avoid waste of time and effort, this, like any other job, needs careful organisation. We need to plan for the maximum of straight runs and the minimum of joinings. It is essential that we co-operate with the bricklayer, and at the same time cover as much ground as possible at each application. The bricklayer should be asked to have as much wall length at D.P.C. level as possible ready at the same time.

As a safeguard against unnecessary damage to the lead, the walls should be carefully inspected and swept off, any lumps of hard mortar must be cleaned off and any sharp pieces of brick removed.

When the bricklayer has laid the mortar bed, the longest lengths of lead should be rolled out quickly in turn—with such speed that the mortar will not have time to dry out and harden. In order to maintain a fairly even thickness of bed-joint, it is best to dress the lead down gently, using a short piece of flooring-board and a hammer. The shorter pieces on sleeper and party walls, and the bay-window and chimney foundations, can now be laid. This order of procedure should minimise the joints needed.

As bituminous-felt D.P.C. is merely overlapped for joining, it can well be argued that this method will be satisfactory in the case of metal D.P.C. However, there is little doubt that a more efficient and reliable joint is justified by the small additional cost of skilled labour.

The development of the method of jointing generally used—the single-lap welt—is illustrated in Fig. 9. Stage 1 shows the ends of the rolls in position, one is turned up about 25 mm (1 in), the other about 50 mm (2 in). The turning down of the overlap is indicated by the dotted lines. The second subsequent turning of the whole welt is shown in Stage 2.

For greater security, it is sometimes recommended that a "solder wiping" should be made on the open edge of the welt. Once all the welts are made, little time is needed to make a large number of wipings with the help of a blow-lamp.

Where two pieces of lead intersect at an angle, the only fully efficient welt is one which is made diagonally across the corner, as shown in Fig. 8. Any other method would necessitate cutting into the 115 mm ($4\frac{1}{2}$ in) sheet to a distance of 25 mm (1 in), thereby reducing the effective width. For the same reason a special jointing piece will be needed where a tee-junction occurs.

Welding Lead D.P.C.

Lead-welding or welding lead is more fully dealt with in Chapter VI. It should, however, be mentioned that lead-welded seams can take the place of welts and soldering. And it is quite

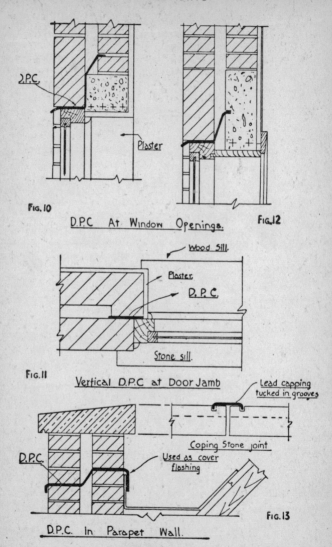

D.P.C.

Plaster

Fig. 10   Fig. 12

D.P.C. At Window Openings.

Wood Sill.

Plaster.

D. P. C.

Stone sill.

Fig. 11

Vertical D.P.C at Door Jamb

Lead capping tucked in grooves

Coping Stone joint

D.P.C.

Used as cover flashing

Fig. 13

D.P.C. In Parapet Wall.

economical in practice if the size of the job justifies the transportation of the welding equipment.

A lapped joint is recommended and is shown in Fig. 33. It is on larger jobs with walls of greater thickness and complexity that lead-burning really comes into its own, by the economy effected in the bossing of special pieces. This is particularly true where the D.P.C. is carried through the whole thickness of the wall, and is stepped up a course on the inside, as in Fig. 13.

Where welts are to be used in the case mentioned above, it should be noted that the welts will have to be made on a flat surface, and the welt subsequently dressed to the shape of the wall.

Damp-proof Courses over Window and Door Openings

It should be remembered that the outside or weathering skin of a cavity wall might be saturated with moisture in wet weather. At no point should this external 115 mm ($4\frac{1}{2}$ in) wall be allowed to come into direct or indirect contact with the inside part of the wall. This calls for special precaution wherever the cavity is enclosed at the head of any window or door opening. Alternative methods of providing a D.P.C. are shown in Figs. 10 and 12. In Fig. 10, we have a brick arch and a concrete lintel abutting, the D.P.C. is so arranged that there is no possibility of moisture passing from the outside to the inside. Where, as in Fig. 12, the concrete lintel is of unusual depth, it might be of advantage to have a groove provided in the casting, in order to economise in lead. Alternatively, in this case, the lead would be carried up to the first bed-joint above.

A D.P.C. is also needed at the side of a window or door opening, where the cavity is enclosed. This is illustrated in the plan in Fig. 11.

Parapet Wall D.P.C.

As rising dampness takes place by capillary attraction against the force of gravity, it will be obvious that downward penetration —aided by gravity—will be much more rapid. In this connection, parapet walls and chimneys are the prime sources of trouble. The position of the D.P.C. in a parapet wall is shown in Fig. 13. The detail shows the method of weathering the joints between the coping stones. The grooves into which the lead is tucked, should be an inch deep, and the lead flashing should either be caulked with molten lead or pointed with linseed oil mastic. Failure to provide a D.P.C. in a parapet wall is a very common source of dampness.

D.P.C. in a Chimney-stack

There has been an increasing tendency in recent years to provide a D.P.C. in large chimney-stacks. Faulty flaunching around the chimney-pots—especially when the area is large—inevitably leads to downward moisture penetration. The D.P.C. is designed to prevent penetration below roof line.

Many forms of D.P.C. have been tried, but everything considered, the arrangement depicted in Fig. 14 proves most efficient and most economical in labour and material. Where the flashing is at apron level, it forms a tray with three of its sides turned up to a height of 38 mm ($1\frac{1}{2}$ in). The fourth side turns to form the apron flashing. (See Chapter V, which deals with chimney flashing.) Inside each of the flues the lead is turned up 38 mm ($1\frac{1}{2}$ in). The complete D.P.C. is shown in Fig. 16. It will be seen to form a tray from which any moisture must run away at the apron, and so down the roof. Where the flashing is at back gutter level, the sides turn down on the face of the brickwork (Fig. 14 (A)).

Practical problems present themselves in the fixing of the D.P.C. Firstly it is essential that the plumber and the bricklayer attend to each other's requirements. Where the chimney is large in cross-section, the flashing will be best made in two or more pieces and joined by means of welts. Having decided into how many pieces the flashing shall be divided, the steps are as follows: (1) work the upturn to the flues, taking care that the openings are not oversize; (2) set up battens and slates so that the upstand of the apron can be measured; (3) work the apron to approximate size; it will not be possible to finish the apron until the slates are in place; (4) place the lead in position.

The rest of the work must now wait until the chimney—or at least a substantial amount of it—has been built. Then the upstand on the three sides can be dealt with. The turning-up of the sides is a simple matter until we come to the corners, when it must be realised that the strokes used in bossing will stretch the lead on the sharp corner of the brick unless great care is taken. Before beginning, it will be a wise precaution to take a hammer and chisel and slightly round off the corner.

In general, what is known as "pig-earing" is regarded as bad practice. Nevertheless, there are occasions when it is a justifiable

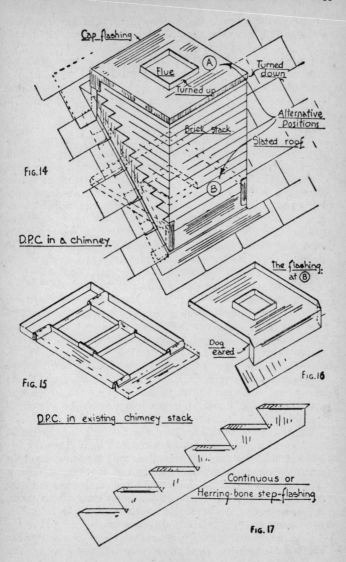

Cap flashing

Flue

A

Turned up

Turned down

Alternative Positions

Brick stack

Slated roof

B

Fig. 14

D.P.C. in a chimney

Fig. 15

the flashing at B

Dog eared

Fig. 16

D.P.C. in existing chimney stack

Continuous or Herring-bone step-flashing

Fig. 17

and advisable practice—as the lesser of two evils. To obviate an opening of the pig-earing with time, double is better than single folding, i.e. half on each side (Fig. 72 (A)).

The placing of the D.P.C. at apron level will not be satisfactory if the roof space is to be used and kept free of dampness. In this case, the D.P.C. is placed at back-gutter level (Fig. 14 (B)), or better still a D.P.C. is inserted at each level.

D.P.C. in Existing Chimney

It sometimes happens that a D.P.C. has to be inserted in an existing chimney. The method is illustrated in Fig. 15. The D.P.C. is laid in a number of pieces by the removal of sections of brickwork and the use of vertical welts, which are arranged to stand in vertical brick joints. On completion, the sides should be turned up at least 38 mm ($1\frac{1}{2}$ in) on the face of the brickwork if it is below back-gutter level. Otherwise it can be trimmed off at the face of the brickwork. In practice, it has been found that the upturn of the lead on the outside of the chimney and on the inside of the flue remains permanently in shape.

BS6 sheet lead is recommended for this purpose, but in practice BS4 sheet lead has been found to be quite satisfactory. It is often claimed that lime mortars and cement mortars have a deleterious effect on sheet lead. There is little evidence to show that excessive damage occurs, but as a precaution a bituminous coating can well be applied. Bituminous paint is readily available.

Sheet-copper Damp-proof Courses

Sheet copper is rapidly gaining in popularity for use in D.P.C.s. Chapter VII deals in detail with the techniques peculiar to the working of sheet copper. Like lead, copper can be jointed by means of welts; it can be soldered and welded.

Copper has many advantages; it is available in appropriate widths as "strip"; it resists corrosion; it is fireproof; it has good ductility, enabling the craftsman to bend or crank the D.P.C. at changes in level; it adjusts itself to the mortar bed without damage. Strip-copper is available in long lengths; this is useful on the longer buildings and sometimes in jointing.

Copper has one important advantage over some materials—it is not acted upon by lime and cement in mortar.

Copper of 24 S.W.G. (0·559 mm) is generally used and need not be exceeded.

CHAPTER IV

LEAD-COVERED FLAT ROOFS, GUTTERS, AND CESSPOOLS

THE method of weathering a lead flat can best be learnt by considering the procedure to be followed in the case of a roof such as is shown in Fig. 18.

Instructions to Carpenter

It is of first importance that the carpenter should co-operate in the provision of adequate support and a satisfactory layout for the lead. We must decide on the falls we need and the depth and position of the drips and rolls.

Many troubles can arise from faulty woodwork. All nails should be well punched below the surface, the boards should be well cramped up and should run in the direction of the fall. The carpenter should be asked to remove the sharp edges of the drips, and should be requested either to fasten down the rolls as the work progresses or to leave them loose for fixing by the plumber.

As shown in Fig. 25, the carpenter should provide a rebate or sinking along the edge of the drip, so that the undercloak can be finished flush with the surface of the boards; so as not to form a ridge.

Having now persuaded the carpenter to provide satisfactory woodwork for the flat, the work of covering can go ahead.

Measuring for Sizes of Lead

Over-generous allowances are often made and much lead is wasted, for fear the sheets should be short in any direction. This is due to lack of confidence arising from uncertainty. As a result of such inaccuracy, there is a continuous cutting away of surplus lead, and, what is worse, there is increased difficulty of handling, and bossing at rolls and drips is complicated.

The sheets should be measured to the neat size of flat surface, and the following allowances should be made for 50 mm (2 in rolls and drips:

(1) Each roll undercloak, a turn-up of 90 mm ($3\frac{1}{2}$ in).
(2) Each roll overcloak, a turn-up of 180 mm (7 in).

(3) Each drip undercloak, a turn-up of 76 mm (3 in).

(4) Each drip overcloak, a turn-down of 100 mm (4 in).

(5) The turn-up to the wall should not be less than 100 mm, and only in exceptional circumstances should 100 mm be exceeded. If 100 mm is exceeded, some method of support should be used, otherwise it will be liable to sag.

(6) The turn-down into the gutter need not exceed 76 mm.

Setting Out

It is when we begin to set out the flat that we realise the need to understand just how lead will behave when it is exposed to the weather. Plumbing students can well be reminded here that scientific principles have application in practical situations and should be understood.

We must know to what temperatures the lead will be subjected; the temperature variation is of greatest importance, because this will determine the amount of expansion and contraction likely to take place. The depth of drips and height of rolls has to be determined so that capillarity can be countered.

Materials exposed to the weather in England are subjected to a great many changes of temperature. It has been said that from this point of view, we have thirty or forty winters each year compared with one on the Continent. Where we have thirty or forty freezes and thaws, the Continental countries generally have one freeze-up at the beginning of winter and a thaw in the spring. In consequence of this, lead on a flat roof in this country will expand and contract a great many times through the rapidity of temperature changes. This is significant because of the peculiar effect of expansion on lead.

Common sense tells us that lead will expand and contract at equal rates, but experience shows that movement due to expansion is much greater than that due to contraction. This is the effect we call "creep." This damaging result of expansion generally takes a downward direction, because the weight of the lead assists expansion and resists contraction. It is almost as though the lead moves on a ratchet gear.

The result of this action is a local stretching of the lead, ending usually in a crack or split.

In order to minimise this effect, it has been found—through many years of experiment—that a piece of sheet lead should not exceed about 2·2 m² (24 ft²), 2·4 m length, or 1 m width. At the

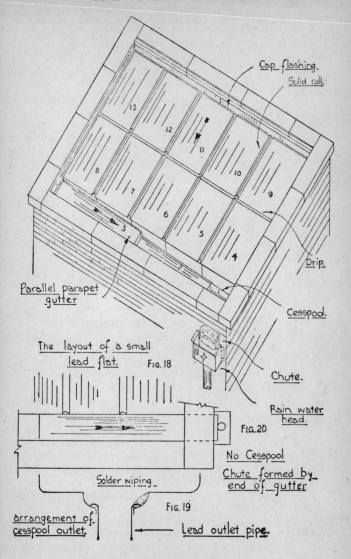

Cap flashing.

Solid roll.

13

12

11

8   10

7   9

6

3   5

4   Drip.

2

Parallel parapet
gutter

Cesspool.

The layout of a small
lead flat.   FIG. 18

Chute.

Rain water
head.

FIG. 20

No Cesspool
Chute formed by
end of gutter

Solder wiping

FIG. 19

arrangement of
cesspool outlet.

Lead outlet pipe.

same time, some means has to be devised for joining adjacent sheets of lead without making them continuous. The problem is one of separating the sheets and at the same time to provide a watertight joint.

This is achieved by means of drips and rolls, as shown in Figs. 22 and 23 and as arranged in Fig. 18.

The drips form a series of steps and run across the direction of the fall. They should be not less than 50 mm in height, and need not be more. Three alternative methods of treating a drip can be considered. The first and most usual has the overcloak finishing over the drip and extending an inch on the flat (Fig. 22). The second method has the overcloak ending about half an inch from the flat. This method should be used where there are heavy soot deposits, because in this situation the first method may aid damp penetration through the space between the sheets on the flat becoming filled with soot, which assists capillarity. Lastly the drip can be fitted with a vee-groove on the upstand, into which the undercloak can be worked. This is known as a capillary groove. Most authorities recommend its use, but it is seldom found in practice—partly because this adds to the cost and partly because of the difficulty experienced in getting the carpenter to provide this "unwonted luxury."

The rolls, which run at right angles to the drips and in the direction of the fall, accommodate the length of the sheets. They, too, must not be less than 50 mm in height and should be cut to the shape of the section shown in Fig. 23. If the roll is worked out of 50 mm (2 in) stuff, it will measure less and finish considerably undersize. Care should be taken to see that the proper finished size is provided, and small, round rolls—like brush handles—should be rejected.

Fixing Procedure

In the example shown in Fig. 18, where we have a lead-covered flat draining into a gutter which empties into a cesspool, the first step will be the lining of the cesspool by bossing or by lead-burning (Figs. 44–48), or even wiped, soldered seams. This will be followed by the lining of the gutter (Fig. 21). The natural order of fixing the sheets will be as indicated by the numbers on the plan; that is, from the cesspool to the left and from the gutter towards the back wall. In certain circumstances, this can be varied; as, for example, when there is a strong, prevailing wind

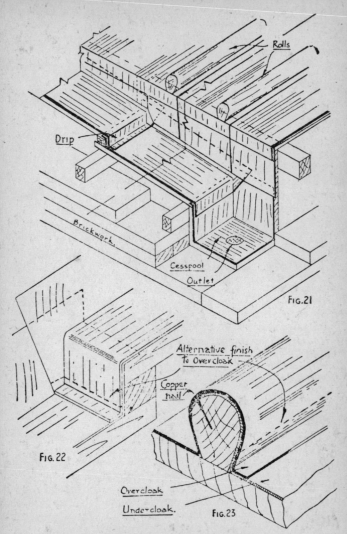

Rolls

Drip

Brickwork.

Cesspool

Outlet

Fig.21

Alternative finish
To Overcloak

Copper
nail

Fig. 22

Overcloak
Undercloak.   Fig. 23

FINISH OF LEAD FLAT AT PARAPET WALL GUTTER.

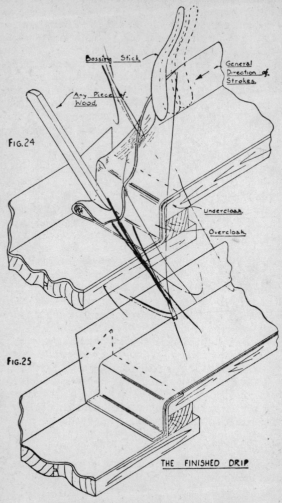

Bossing Stick

Any Piece of Wood.

General Direction of Strokes.

FIG. 24

Undercloak.

Overcloak.

FIG. 25

THE FINISHED DRIP

METHOD OF WORKING OVERCLOAK.

and driving rain from right to left (on the plan), then it will be advisable to begin on the left.

In the interests of time and economy of energy, it will be advisable to fix up a temporary bench—on the flat roof if necessary—on which the pieces of lead can be worked. Two planks, resting on boxes or trestles, will be found to be satisfactory.

Having set out the positions of the rolls, the first piece of lead should be measured to the allowances already mentioned and the corners should be worked in readiness for placing the sheet in position. The corners where the sheet meets the drip and roll, should be bossed to a height of 50 mm, and the corner worked down to the roll and into the rebate on the edge of the drip *in situ*. The first sheet having been prepared should be placed roughly in position against the wall and up to the drip. It should now be drawn back about 50 mm and "bumped up" two or three times, this ensures a good fit in the angle. The first roll should now be placed in position and driven hard up against the upstanding undercloak and screwed down; this will form a good angle. The undercloak is now dressed over and the ends worked down.

The second and successive sheets are similarly fitted, and the rolls screwed down until the last piece is reached. This sheet will need rather more careful measurement and setting up, as its limits are decided by the position of the last roll and the wall.

It may happen that there is a wide turn-down into the gutter; particularly at the cesspool end. In this case copper straps of stout quality must be previously fixed and folded over the end of the overlap. This will serve to hold the lead down in position.

In bossing the ends of drips, excessive working can be avoided if a method such as is shown in Fig. 24 is used. There the overcloak is almost placed in position and the lead is stretched where it is of least importance; that is, under the cover flashing. The alternative method so commonly used stretches the lead where it is exposed and weakens its weathering qualities.

The Gutter Outlet

The gutter outlet may be dealt with in various ways. Three methods are shown in Figs. 18, 19, and 20.

CHAPTER V

LEAD FLASHINGS

FLASHING is a term used to describe the process of weathering the intersection between, say, a chimney and the roof it penetrates, or where the slates meet at a ridge or in a valley gutter, or again where a ventilation pipe passes through a slated or tiled roof.

Flashing a Chimney-stack

Just as slates must overlap to ensure that the rain-water will be carried down and off the roof, any lead flashings used on a chimney must be laid in the same way.

The Apron (Fig. 29)

As the name implies, this flashing is fixed to the front face of the chimney, from where it carries the water on to the slates below. In calculating the size of sheet lead needed for an apron, measure the width of the chimney and add 150 mm (6 in) at each end for turning the corners. For depth, 150 mm is allowed to lie on the slates and the upstand is measured to the second nearest horizontal brickwork joint. This will vary from chimney to chimney. An allowance of 25 mm (1 in) is also made for turning into the horizontal joint. It is not uncommon for this allowance to be reduced, to save the trouble of cutting out the mortar to the depth of 25 mm. This is bad practice, because so much depends on a successful joint with the brickwork. The ends of the apron are turned by driving the lead in waves radially from the front to the ends of the 150 mm wide part, which will be on the slates. This operation is best carried out on the bench after careful measurements have been taken along with the pitch of the roof.

Cutting out the Brick Joints

It is most satisfactory if the bricklayer can be persuaded to rake out the joints to the desired depth before the mortar hardens, especially if cement mortar is being used. The plumber is well advised to be on hand as the chimney-stack goes up, to do this himself or mark out the joint to be cleaned. If the mortar has set

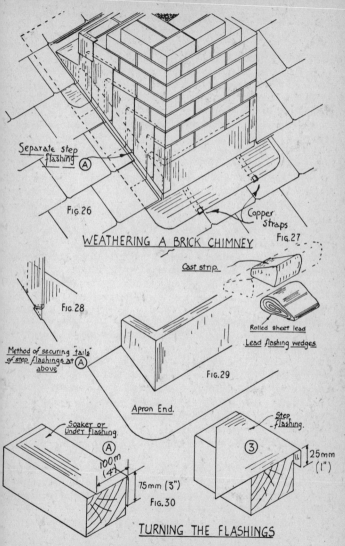

Separate step flashing Ⓐ

FIG. 26

Copper straps

FIG. 27

WEATHERING A BRICK CHIMNEY

FIG. 28

Cast strip.

Rolled sheet lead

Lead flashing wedges

Method of securing "tails" of step flashings at Ⓐ above

FIG. 29

Apron End.

Soaker or Under flashing.

Ⓐ

100m (4")

75mm (3")

FIG. 30

Step flashing.

③

25mm (1")

TURNING THE FLASHINGS

a plugging tool, or raking chisel, should be used, because by its shape it will neither stick nor damage the brickwork.

The Back Gutter and Soakers

On the sides of the chimney a soaker flashing is inserted at every course of slates. This is a piece of lead which is folded (Fig. 30 (A)) to a right angle so that 100 mm (4 in) lies under the slate whilst 75 mm (3 in) forms an upstand against the brickwork. The soakers are shown in position by means of broken lines in Fig. 26. The first soaker overlaps the end of the apron and is overlapped by the next soaker, and so on to the back of the stack, where the last soaker is overlapped by the end of the back gutter, thus forming a continuous weathering.

To save loss of time through hanging about, it is usual to hand the already turned soakers to the slater for fixing as the courses of slate or tile are laid, and each soaker is itself nailed with the course, otherwise the soakers might slip and the overlap be lost, with consequent "raining in."

The method of calculating the length of the soaker must be considered; generally a formula is given as follows:

$$\frac{\text{Length of slate} - \text{lap}}{2} + \text{lap} + 50 \text{ mm}$$

This is used because slates of varying length are used. Lap is usually taken as 75 mm, and the 25 mm addition is an allowance for nailing.

In practice, the plumber measures the margin (the distance between the ends of the slates when fixed) and adds 75 or 100 mm. Nevertheless, the formula is useful where it is convenient to cut the soakers when the slating has not begun.

The soaker is fixed so that its lower edge coincides with the end of the slate it is weathering.

The Back Gutter

This flashing (to be seen in Fig. 26 and Chapter VI) weathers the back of the chimney. The lead is supported by a wood gutter, provided and fixed by the carpenter; this consists of a sole, usually 150 mm wide, and a board running some distance up the roof.

In calculating the size of lead to be used, allow as follows:

In length, allow the width of the chimney plus 150–200 mm

(6–8 in) at each end, depending on the pitch of the roof. A steep roof requires more lead.

The width requires about 125 mm (5 in) of upstand, depending on the position of a convenient joint, with 25 mm to turn in. Add to this 150 mm for the sole and 200–250 mm (8–10 in) to lie up the roof under the slates. Altogether, about 500 mm (20 in) is usually needed.

Cover Flashings

Having made and fixed the apron, soakers, and the back gutters, the rain falling on the roof surface will have been effectively prevented from passing into the building. We need now consider how to deal with rain-water running down the faces of the chimney.

To this end, overlapping cover flashings are provided.

In the case of the apron, the cover flashing is included by turning the upstand into the brickwork joint. Similarly, this may be done in the case of the back gutter, but often a separate strip of lead is turned into the joint and allowed to overlap the upstand by about 75 mm.

This leaves the soakers to be weathered. Two common methods are met with. Figs. 14 and 17 illustrate herringbone-step flashing; here a piece of lead is set out in a number of continuous steps. In this method setting-out is of great importance and accuracy in measuring the pitch of the roof vital. It gives a neat appearance and leaves no ends for the wind to lift up.

Fig. 26 shows the common alternative, found mostly in the north. Here separate and overlapping steps are formed; they are turned into the brickwork a minimum of 25 mm; they overlap the soakers by 50 mm, finishing clear of the slates so that debris will be less likely to accumulate, and cover each other by 40–50 mm ($1\frac{1}{2}$–2 in).

It is especially important that the pointed tails of the flashings be properly secured. The method is shown in Fig. 28. This precaution should not be neglected, as chimneys are generally subject to gusty winds and lashing rain.

The apron, too, should be supported at its lower edge. This is best done by means of one or two strong copper straps.

Lead wedges should be used to fasten the flashings into the brick joints. Fig. 26 indicates two kinds. For an occasional job a piece of sheet lead, rolled and flattened as shown, can be used.

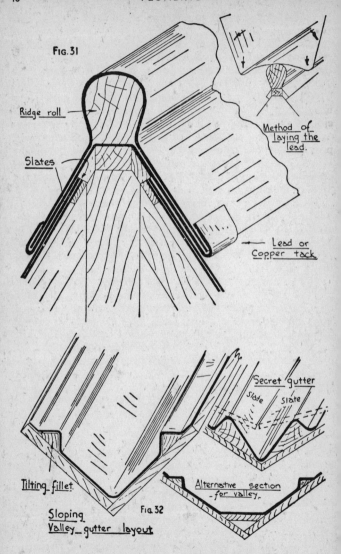

FIG. 31

Ridge roll

Slates

Method of laying the lead.

Lead or Copper tack

Secret gutter
slate   slate

Tilting fillet.

Alternative section for valley

Sloping Valley gutter layout

FIG 32

But it is useful to have a mould from which strips of wedge can be cast, to be cut up into small pieces as needed. These wedges should be well caulked into position, by means of a blunt-edged drift and not a sharp chisel. Great care should be taken to see that the edge of the flashing does not get cut or be torn in the process.

When the flashing has been completed, the brickwork joints into which lead has been turned must be carefully pointed, either with the mortar used in the brickwork or with oil mastic.

When flashing chimneys, use a "cat ladder" for safety and protection of the slates or tiles.

Figs. 31 and 32 illustrate the flashing of ridge rolls and valley gutters of varying type. The method of forming the ridge roll flashing is shown. Proper fastenings or tacks are important, because of the exposure to wind to which a ridge is subjected.

Valley gutters are formed as shown, and the sheets overlap each other. They should be securely nailed behind the tilting fillet.

CHAPTER VI

LEAD-WELDING

LEAD-WELDING is the modern method of autogenous or fusion welding of lead, i.e. joining of one piece of lead to another with the help—if necessary—of a "filler-rod" of lead. This as distinct from soldering, where an alloy is used to join the pieces of lead.

Reasons for Lead-welding

Until very recently this technique was used mainly in the chemical industry. Here it was found that solder could not be used for jointing lead pipes, or on the lining of vats and other containers, because the tin in solder was found to react with the chemicals and a failure of the jointing resulted.

These facts have been known for centuries. In the early days of lead-welding from many centuries ago until the development of combustible gases in the early part of the nineteenth century, lead-welding was carried out by a simple process of pouring very hot molten lead into a vee-joint formed between two sheets of lead—the underside being packed with sand, which prevented the molten lead running through. In vertical positions this technique was used by placing an iron mould on the face of the joint.

The introduction of combustible gases has made it possible to carry out lead-welding operations in almost any situation with joints on sheets and pipes in almost any position.

The easy availability of gases in transportable cylinders has made lead-welding an economic proposition for building work and even in domestic plumbing. Lead-welding is particularly suitable where work can be prefabricated; for example, where a large number of lead-slate flashings can be made for a housing scheme or a large number of unions can be burnt-on to lead flush pipes. These are just two of many possible prefabrications.

Lead-welding Gas Systems

There are several gas systems available for lead-welding.

Air-Hydrogen.—This system was the first, and was made

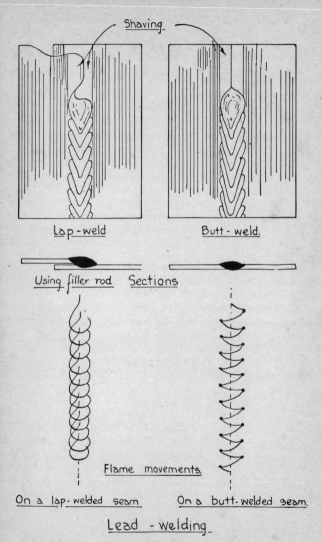

Shaving

Lap - weld

Butt - weld.

Using filler rod   Sections

Flame movements.

On a lap-welded seam   On a butt-welded seam.

Lead - welding

possible by the invention of the air-hydrogen blow-pipe something over a century ago. Hydrogen was produced by the reaction of sulphuric acid with zinc spelter in a specially constructed lead-lined machine. Air at a constant pressure was made available by means of a bellows and a weighted storage vessel of the gasometer type. This system was in fairly common use twenty to thirty years ago.

Oxy-hydrogen.—This system is widely used in chemical works. The gases are available in cylinders. An injector-type blow-pipe is used, and controlled by means of pressure valves and gauges. The flame temperature is about 2,300° C.

Oxy-coal-gas.—Although coal-gas can be obtained in cylinders at high pressure, the advantage in this system lies in the fact that coal-gas is so widely available in mains.

If mains gas is used, an injector blow-pipe is required to increase the velocity and therefore the quantity of gas from the main. Town's gas usually has a pressure of about 70 mb, and this pressure is insufficient for the normal blow-pipe. Flame temperature about 2,000° C.

Oxy-acetylene.—This is the system now most universally used. The flame temperature (3,500° C) is the highest of any of the gas combinations usually available, and the cylinder gauge and blow-pipe adjustments can be so finely made that lead-welding can be carried out in any situation without great difficulty by skilled men. This system is useful, too, because it can be used so readily for the welding of all other metals.

The manufacturers of welding equipment produce a special blow-pipe (No. 0) and five interchangeable tips for use with oxy-acetylene.

Selecting the Right Tip

It is important that the right size of tip should be used for each particular job, depending on the thickness of the lead and on the burning position. The table on page 48 gives an indication of those generally most useful, but it should be noted that tip size also depends on the skill and speed of the lead-welder. At increased speed, larger sizes can be used.

The Flame

According to the proportions of oxygen and acetylene used, three kinds of flame can be produced. When the gases are in

SUITABLE TIP SIZES

Weight of Lead		Size of Tip				
		Flat	Vertical	Overhead	Underhand	Horizontal
kg	lb					
1·36	3	1 or 2	1	1	1	1
1·8	4	3	2	2	2	2
2·3	5	3	2	2	2	2
2·7	6	4	3	2	2	2
3·2	7	4	3	3	3	3
3·6	8	5	3	3	3	3

WORKING PRESSURES WITH VARIOUS TIPS

Size of Tip . .	No. 1	2	3	4	5
	bar gauge pressure				
Working pressure (both gases)	0·1	0·17	0·2	0·275	0·35

equal proportion, we get a neutral flame, this is recognisable by the clearly defined bright blue inner cone at the tip of the torch (see Fig. 79). The detail (B) indicates the shape the cone ought to be, its end should be rounded. This is the flame used in lead-burning. If there is an excess of oxygen, the flame will have a very small "thin" and pointed inner cone and the whole flame will be reduced in size. This is known as an oxydising flame, it is not suitable for lead-welding as it prevents satisfactory fusion by oxydising the lead. If on the other hand we have an excess of acetylene, the flame is described as a carbonising or carburising flame. In this case (detail A) the inner cone is seen to have a feathery end. When a flame of this kind is used, it is found to be "dirty," and the lead takes on a black coating of carbon, which is also a hindrance to fusion.

Preparation of Seams

Lead-welded seams are either butted or lapped, depending on the particular job, the strength required, and the burning

Apron flashing.

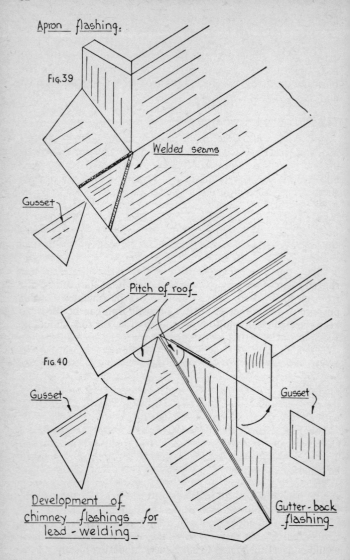

Fig. 39

Welded seams

Gusset

Pitch of roof

Fig. 40

Gusset

Gusset

Development of chimney flashings for lead-welding.

Gutter-back flashing

position. Butt joints will be used where a flat surface is needed, or where a strong joint is unimportant. A butt-joint should be prepared so that the edges fit snugly together and no gaps exist. This precaution is particularly important where the weld is to take place on a wooden surface, for the flame might burn the wood and the gases produced tend to bubble through and make the weld porous.

The Butt-weld

In making a butt-weld, it is necessary to use a filler-rod; that is, a strip of lead which can be melted on to the joint to increase its thickness. This build-up of thickness is essential in welding milled sheet lead if the joining is to be as strong as the original sheet. Milled sheet lead has a tensile strength of over 20×10^6 N/m² whilst cast lead has a tensile strength of not more than 14×10^6 N/m². When milled lead has been melted, it becomes cast lead, so that it will be obvious that an increased thickness is needed to make a weld as strong as the rest of the sheet.

Filler-rods can be made from strips cut from a sheet and shaved clean of the grey oxide which coats it, or they can be cast from clean pig lead. Special moulds can be obtained for this purpose or the corner of a piece of angle-iron can be used. To keep the lead clean and to prevent oxidation, the rods should be wrapped up or buried in sawdust.

The build-up of the butt-weld can be seen in Fig. 34. Here it will be noticed that the seam is made by means of a number of overlapping pools of molten lead. These are contrived by a flame movement, as illustrated in Fig. 38. This movement ensures that the pool will fuse with each sheet, whilst succeeding pools merge with each other and form a continuous weld.

The illustrations show that each piece of lead has been shaved clean of oxide for a distance equal to half the width of the seam. This preparation must not be neglected if sound results are to be achieved.

The section through the butt-weld shows how the burning should penetrate the full thickness of the sheet. For this reason, it is important that the edge of the sheet should be shaved.

Grease should not be used to prevent further oxidisation of the cleaned parts, as is done in soldering, because the intense heat would convert the grease to substances hindering fusion.

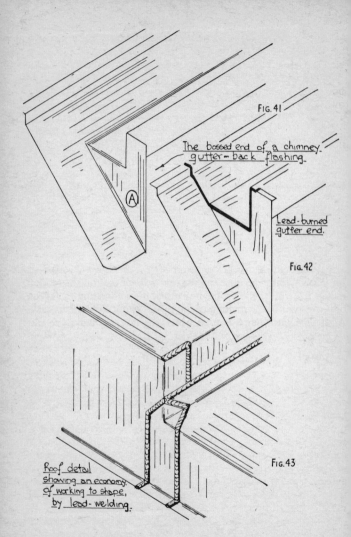

FIG. 41

The bossed end of a chimney gutter-back flashing

Lead-burned gutter end.

A

FIG. 42

FIG. 43

Roof detail showing an economy of working to shape, by lead-welding.

The Lap-weld

The lap-weld (Figs. 33 and 35) is used where greater strength and security is needed, or where the seam is so placed that it would be difficult to make a butt-weld; e.g. in upright positions.

The flame movement (Fig. 37), sometimes called the "button-hook," produces a continuous weld by melting the edge of the overlap and continuously fusing this with the preceding "pool" and the underlapped sheet.

It is most important that both the upper and under surfaces of the overlap should be cleaned as well as the full width of the seam on the underlapped sheet. In this way, no oxidised lead will be involved in the welding process. A general rule might be given: all surfaces which will be melted must be cleaned.

A filler-rod is not required in lap-jointing, unless additional strength is wanted. In this case, a second run is made and additional metal welded on.

Vertical Lap Joints

A very high degree of skill in handling the flame is required for vertical joints because of the gravitational pull on the molten metal. It involves the problem of constantly adjusting the speed of movement to the melting of the lead without loss of accuracy. A small neutral flame is used generally, but the flame size can increase with the development of speed and accuracy. A good rule is to begin with a very small flame, because it gives more time for careful movement.

The procedure is as follows: See that the overlap is lying snugly against the underlap. Begin at the bottom, and by almost simultaneously melting the overlap and underlap, carry a small bead of lead from the edge of the overlap—by a button-hook movement—to form a lip on the underlap. This lip forms a shelf for the support of the next small bead. As this bead is carried down and round, it is at once fused with the lip and the underlap. In repeating this movement, a continuous weld is achieved. If on this seam, or indeed on any lead-welded seam, there should be a break in the continuity of fusion, it will be readily apparent and the movement should be repeated.

It is at such times that difficulty might arise and the seam be spoiled, because in repeating the bead formation at the same point there is an inevitable tendency to creep into the lap and

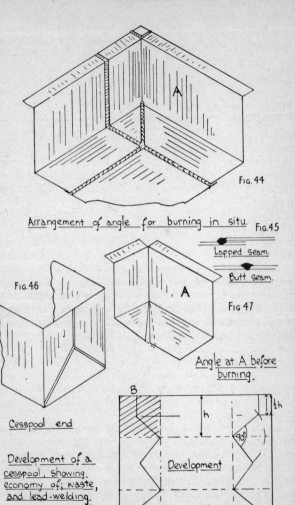

FIG. 44

Arrangement of angle for burning in situ.   FIG.45

Lapped seam.

Butt seam.

FIG. 46

A

FIG 47

Angle at A before burning.

Cesspool end

Development of a cesspool, showing economy of; waste, and lead-welding.

FIG. 48

Development

move the seam out of vertical. Once the seam begins to creep it is difficult to recover; so that unless considerable skill has been developed, it will be as well to introduce a bead from a filler-rod.

It will be obvious that in lead-welding of this kind great steadiness and accuracy are essential. Having grasped the fundamental principles, the beginner must do a lot of practice.

Horizontal Lap Joints

Horizontal joints are less difficult to manage than those in vertical positions, but they have their own difficulty. This is the tendency of the flame to cut into the upper edge of the seam, reducing the effective thickness of the sheet lead. To overcome this tendency, the overlap which presents an upward edge is turned outwards to form a lip, and the seam is built up by using a filler-rod. This is the overhand joint in Fig. 51.

The alternative horizontal joint is the underhand joint (Fig. 54). This seam is made in much the same way as the vertical lap-joint, but where possible it is to be avoided in favour of the overhand method.

Other lead-welded seams are shown in Figs. 49–53, as used on a lead-slate flashing and at angles. There are, of course, many situations to be met with where a seam is a combination of the types mentioned. For instance, in welding or "burning-on" a circular patch of any size the seam will be partly left-, and partly right-handed vertical lap; at the top it will involve overhand burning and the bottom underhand burning.

The Application of Lead-welding

The lead-slate flashing is taken as an example of the use of lead-welding as an alternative to the bossing method. Jointing details are outlined in Fig. 50.

As an alternative to cutting out the opening in the sheet, the upstand is sometimes welded on, and the opening made by cutting out the ellipse within the upstand.

The method of developing an apron and back gutter is shown in Figs. 39, 40, and 42 respectively. The most economical method with a gutter-back is that in Fig. 40, but when odd pieces of lead are to be used up, other alternatives are used.

An example of the economical use of lead-welding is illustrated in Fig. 43. Here a few judiciously arranged seams can save a great deal of bossing with considerable saving of time.

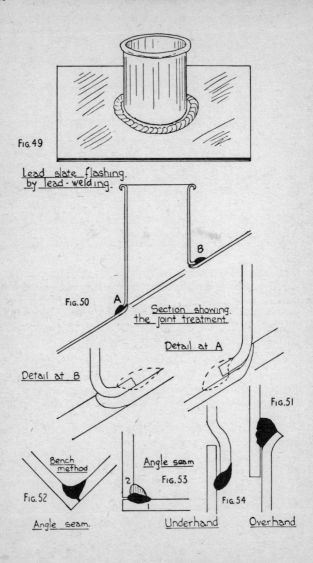

Fig. 49

Lead slate flashing
by lead-welding.

Fig. 50   A

Section showing
the joint treatment

Detail at A

Detail at B

Fig. 51

Bench
method

Angle seam

Fig. 53

Fig. 52

Fig. 54

Angle seam.

Underhand

Overhand

A box-lining or cesspool is dealt with in Figs. 44–48. In Fig. 48 the development of the method used in Fig. 46 is superimposed on the layout where angular seams are used. The saving in the size of sheet needed and in the total length of welding are readily apparent.

Fig. 44 is a method where it is arranged to burn as many seams as possible on the flat.

Welded Joints on Lead Pipes

The techniques discussed above are applied to the lead-welding of joints on lead pipes.

The pipe ends are prepared in the same way as sheet lead, to provide a good fit and a surface cleared of lead oxide.

It is usual to build up pipe joints in stages to ensure a sound weld. In the case of a joint on a waste pipe, two-stage welding will suffice; but where a joint of greater strength is needed, as on a water pipe, it is best to use a built-up joint.

In this method, the socket is opened out to form a wide flange and the spigot is welded on by means of a seam at the base of the socket. Now the socket is partially closed and a second seam is made. This process is repeated until the socket is closed and the joint is completed.

CHAPTER VII

SHEET-COPPER ROOF-WORK

SHEET copper has been used for roof covering for a great many years, but with the improvement in the quality of manufactured sheet, the advance in knowledge of manipulation of the material, and perhaps most significantly its relative cheapness, have made it most popular. Plumbers have inevitably had to adjust themselves to change, and it seems likely that for economic reasons copper will very largely replace lead as the common roof covering.

Copper as a roof covering has two main advantages over lead: Firstly, because of its lightness, copper allows a lighter roof construction. By rough comparison, area for area, copper is one-fifth or one-sixth the weight of sheet lead. The second advantage lies in the fact that copper does not "creep," as does lead. This reduces the need for some of the fastening needed by sheet lead, and makes copper sheet suitable and durable on vertical or near-vertical surfaces.

Sheet copper as used for roof-work, is generally hot-rolled sheet of dead-soft temper; this softness ensures that work hardening will not make the copper brittle. Over-working and hammering must be avoided for the same reason.

Sheet copper of 23 and 24 S.W.G., weighing 5·8 and 4·9 kg/m², is most commonly used and is found to be quite strong enough to withstand reasonable foot traffic.

Like lead, copper is almost everlasting. When copper is exposed to the action of the atmosphere, chemical change takes place at the surface; the oxygen and carbonic acid in the air re-acting with the copper to produce a coating of copper oxide and carbonate. This green and handsome "patina" serves to prevent any further depredation by the atmosphere.

Laying Sheet Copper

The minimum fall permissible for a sheet-copper-covered roof is reckoned to be 25 mm in 2 m, or 1 in 80.

As with lead, it is thought desirable to run the under-boarding in the direction of the fall, but in practice, with the usual ridge-

Tools for sheet copper work

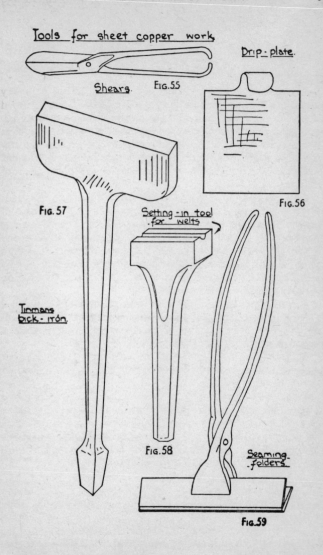

Shears.   FIG. 55

Drip-plate.

FIG. 56

FIG. 57

Tinmans
bick-iron.

Setting-in tool
for welts

FIG. 58

Seaming-
folders.

FIG. 59

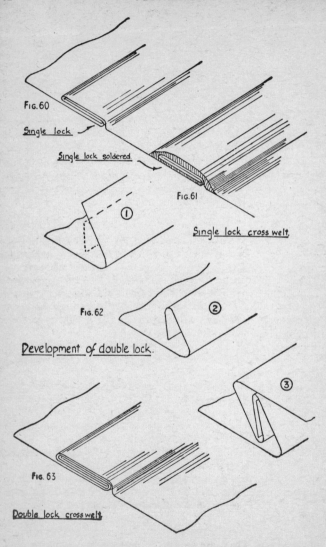

Fig. 60

Single lock

Single lock soldered.

Fig. 61

Single lock cross welt.

① Fig. 62

② Development of double lock.

③ Fig. 63

Double lock cross welt.

and-rafter roof, this is not easily done. The best alternative is to have the boards laid diagonally.

All iron nails must be well punched home because of the possibility of electrolytic action where different metals are in contact in moist conditions. Iron having the lower potential, the nails would tend to deteriorate. A further chemical danger lies in the possibility of deleterious action where copper is in contact with iron rust.

As a precaution, it is thought wise to use an under-layer of building paper; both for the reasons given above and to eliminate resonance due, amongst other things, to spattering rain-drops.

Sheet copper is most easily and economically available in sizes giving a superficial area of 1·3 m²; for instance, 2 m × 0·65 m. In general this area is not exceeded in a single piece of copper, for reasons of expansion and handling.

Cross Welts

Drips are unnecessary in sheet-copper roof covering and are replaced by the use of cross welts. These can be either single lock, single lock soldered, double lock, and double lock soldered. They are illustrated in Figs. 60–63. Fig. 60 shows the single lock and a solder wiping is added in Fig. 61. Soldering, whilst useful in certain difficult situations, is generally to be avoided as unnecessary. A well-made welt will remain watertight even when standing under an inch of water.

For greater security, the double welt (Fig. 63) is to be recommended. The development of this method of cross-welting is shown in three stages in Fig. 62.

In covering a roof surface, it is usual to begin by preparing the narrow ends of the sheets for cross-welting. The finished welt is usually about 12 mm wide and about 75 mm of the length of each sheet is used up in jointing, so that in 2 m sheets the distance between welt centres will be 1·925 m.

When the ends have been prepared, a number of sheets are strung together. In order to avoid the difficulty of joining the ends of two welts on a roll or standing seam, it is advisable to stagger the welts so that they do not meet in adjacent strings. When this is done, a "closer" piece will be needed to bring the ends in line.

The next problem is that of making the joint between the sides of the sheets. There are two main methods, excluding welding.

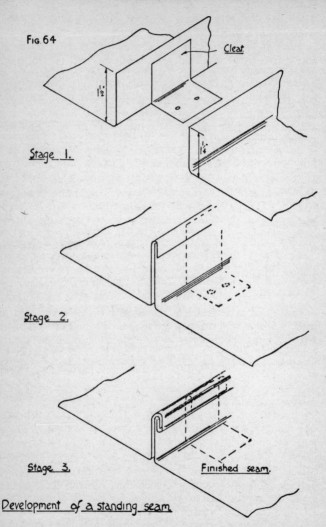

Fig. 64

Cleat

1½

Stage 1.

1¼

Stage 2.

Stage 3.   Finished seam.

Development of a standing seam

They are: by the use of a standing seam (Fig. 64) and, alternatively, using a wood roll (Figs. 65–69).

The standing seam, the development of which is shown in Fig. 64, is a highly satisfactory method of jointing in the direction of the fall. But is not suitable where the roof is exposed to foot traffic. It will be seen that if the seam is flattened it will hardly be effective. This form of seam must be supported at regular intervals by means of stiff copper cleats (Fig. 64, Stage 1); 0·3 m centres being suggested. The cleats must be properly secured with copper nails.

Formation of Standing Seams

The edges of the sheet should be turned up; one to a height of 38 mm; the other to 32 mm, with the cleat standing between them to 38 mm. The extra height of the taller edge is now to be turned as shown. For this purpose a piece of wood 32 mm thick and a heavy dresser are very suitable. The wood is held against the smaller turn-up, and the edge of the taller piece is turned over on to it. Next, the piece of wood is placed against the opposite side and the overlap is dressed down over.

Special Tools

The tools to be found in the usual plumber's kit, such as mallets, shears, and dressers, are used in this kind of work, with the addition of certain special tools, most of which can be improvised. They are pictured in Figs. 55–59. Many others can be devised, of course, and expensive but useful turning clamps can be bought.

Treatment of Wood Rolls

The different types of wood roll are shown in Figs. 65–69. The isometric sketches in Fig. 65 show the development of the finished job in stages. Once more, prepared fillets of wood are needed to help in folding the welt. The most common roll is the conical form. This shape has the advantage that it easily takes up expansion in the sheets at each side. As with the standing seam, the copper has to be secured by means of cleats; the arrangement is indicated in the sketch (Fig. 65).

The ornamental forms of roll are mostly used where the roll is easily seen, and may be used as an architectural feature.

Rolls should always be used where the roof might be walked on.

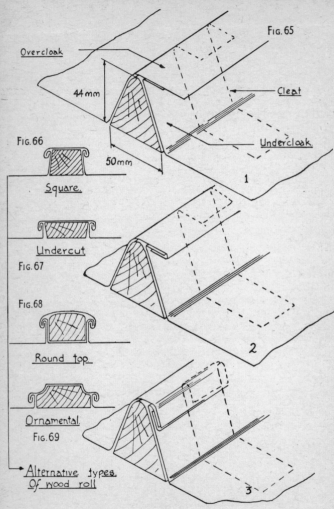

Fig. 65

Overcloak

44 mm

Cleat

Undercloak

50 mm

1

Fig. 66

Square.

Undercut
Fig. 67

Fig. 68

Round top

Ornamental.
Fig. 69

2

Alternative types.
Of wood roll

3

Development of the conical roll.

Bronze-welded Joints

Where oxy-acetylene plant is available, the sheets may be jointed by means of bronze-welded seams. These are described in Chapter VIII, and shown in Figs. 73–76.

Sometimes autogenous welding, i.e. joining copper to copper with copper filler-rod, is used, and a flash-welded seam is the most convenient. (See Figs. 77 and 78.)

The method involved in these processes is more fully explained in Chapter VIII.

Manipulation of Corners, etc.

Where the copper sheet finishes against a wall and a corner has to be contrived, the corner may either be pig-eared or soldered; or where plant is on hand it may be welded.

The treatment of particular angles and corners is almost a matter for individual consideration. Where possible, welting or welding should be used, but sometimes it is extremely difficult to avoid the use of soldering. One precaution should be carefully observed; that is that the flux used should be non-corrosive—of the resinous kind, and should be wiped away when the soldering is finished.

The problem of dealing with the finish of both standing seam and roll against an upstand is shown in Figs. 70–72. In the case of the standing seam a saddle piece is used (Fig. 71). For the roll finish, a capping piece is used.

At an upstand it is advisable to introduce a triangular fillet of wood. This reduces the hardening which results from excessive working, in that more stresses are set up in forming right-angle turns than less acute angles. The effect of expansion is also less marked.

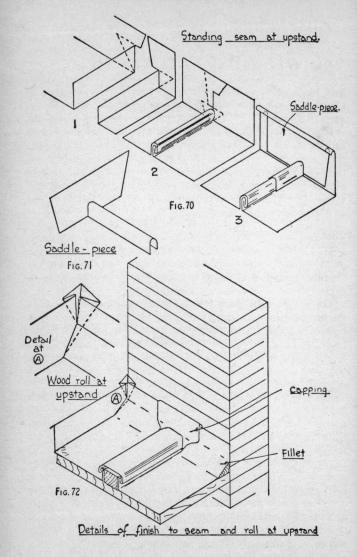

Standing seam at upstand.

Saddle-piece.

1

2

Fig. 70

3

Saddle - piece
Fig. 71

Detail at Ⓐ

Wood roll at upstand
Ⓐ

Capping

Fillet

Fig. 72

Details of finish to seam and roll at upstand

CHAPTER VIII

WELDING AND BRAZING

THE gas systems available for welding have been discussed in Chapter VI, together with the general characteristics of oxy-acetylene flames.

In recent years the plumber's-trade province has widened to include the use of welding techniques on mild-steel pipes, copper sheet and tube, and on lead sheet and pipe. Lead-welding has been fully considered in Chapter VI. Any plumber should be able to manage these techniques, and would indeed be wise to acquire enough skill to be able to carry out simple welding jobs in most other common metals.

Mild-steel Welding

Mild steel presents fewer difficulties than the other metals the plumber may be called upon to weld. In his own terminology—"it runs nicely." It is therefore a good medium with which to begin.

The first essential in any jointing is that of getting a good fit of the surfaces, and time spent to this end is well used. In making a butt-weld on mild-steel piping, the vee-groove provided by the bevel produced by the pipe-cutters ensures a deep penetration in welding. The inside of the pipe should be cleared by reamering. Similarly, a branch joint should be carefully fitted, and not have the pipe end "pushed anyhow into a roughly cut hole."

The first real problem is that of blow-pipe control. Whilst some guidance can be given, the only effective learning will come with practice—provided the learner is both observant and introspective; that is, he should look carefully for the causes of his failures and successes.

The condition and behaviour of the flame must be thoroughly understood. The acetylene should be turned on a little and a light applied to the blow-pipe. A smoky flame will result, more acetylene should now be turned on until the flame burns brightly and the smoke has disappeared. Now the oxygen tap should be opened slowly until a sharp green cone is seen at the tip. As this will be an oxydising flame and undesirable, the oxygen is slightly turned back until a faint fuzziness appears at the tip of the cone. The flame is now neutral and correct for use. Constant inspection of the flame is necessary for a beginner, to make sure that the flame has not changed to oxydising or carburising by a knock or an inadvertent finger movement on the controls. Once the welder has some experience and skill, he will be guided by the behaviour of the flame and its effect on the weld.

In the same way the welder comes to know just how far away from the metal the tip of the cone should be held. It is recommended that the blow-pipe—or torch as it is often called—should be inclined at an angle of 45° to the weld.

As a first exercise in welding, it will be best to begin with a straight weld on sheet-iron, by taking two pieces of about 16 S.W.G. metal, and putting two straight edges together to make a butt-weld.

The blow-pipe is held in the right hand and with the cone near the surface of the two pieces of metal. It is moved slightly from side to side until a pool of molten metal has formed on each edge. A filler-rod held in the left hand is now brought to the tip of the cone—without touching the metal—until a bead of molten metal from the rod falls to unite the pools on the sheets. This movement is now repeated to overlap the previous and still molten portion; and so a weld is formed to give a continuous jointing. The next point to consider is the depth of penetration. This is right when the weld produces a slight bead on the underside and the joining cannot be seen. As the underside of welds cannot always be seen, as in pipe welding, skill must be developed in estimating this from the outside of the weld.

Heating pipes of mild steel are now often welded, particularly in panel-heating systems, where batteries of piping are buried in walls, ceilings, or floors. In the case of a branch joint on, say, a 50 mm (2 in) diameter pipe, welding will save the cutting of three threads and the cost of a tee-fitting; so that economically welding is a good proposition.

It scarcely need be said that as much work as possible should be prefabricated. For example, the heating panels mentioned above should, where possible, be assembled and welded in the workshop or on the floor, and hoisted into position later. When butt-welding the ends of two pipes, it is convenient if the pipes can be rotated, in which case the weld is the same as a flat weld on sheet metal. However, *in situ* welding is not difficult with a little practice, provided the molten pool is kept small, a filler-rod is used, and the flame is softly neutral. The greatest menace is oxidisation, and the oxide once formed must be chipped off. A weld must be made steadily and rhythmically, so that no undue stresses are set up in the metal around the weld. The welding should not be done in parts, but should be continuous; otherwise fractures may occur and result in failure under pressure.

The filler-rod for mild steel is of mild steel, and is copper coated so that it will not readily oxidise. For most pipe-welding a 3 mm ($\frac{1}{8}$ in) rod is suitable.

Every pipe which is to be welded must be properly supported about the weld, so that sagging and distortion is avoided.

Welding Copper Tubes

Autogenous Welding

By autogenous welding we mean the fusion of two pieces of the same metal without the use of any other metal or alloy; lead-burning is autogenous welding of lead, the mild-steel welding technique described above is also autogenous welding.

Although bronze welding—described below—is commonly used, copper welding is not uncommon. There are two main difficulties with copper; firstly, it is a remarkable conductor of heat, and is therefore difficult to raise locally to welding heat. This calls for preheating in many situations and the use of a hotter flame than is usual, say, with mild steel of equivalent thickness. Secondly, copper oxidises excessively. This trouble can be obviated in two ways; the copper can be obtained in de-oxidised form—most copper tube used in building is of this type—and a flux can be used. The flux should be borax based, with the addition of other de-oxidising agents. Fluxes are sold ready prepared and are excellent in quality. Similarly special filler-rods are manufactured, these contain de-oxidising agents such as silicon and phosphorous.

Autogenous copper welding has two principal methods: the "flash" method and using filler-rod.

In flash welding (Fig. 77) the edges of the sheets are raised, as shown, to a height of about twice the thickness of the metal. In the same way the ends of two pipes are flanged for welding. Now, by using a rapid circular movement, the edges are fused together. The flame must be neutral, and should be held just

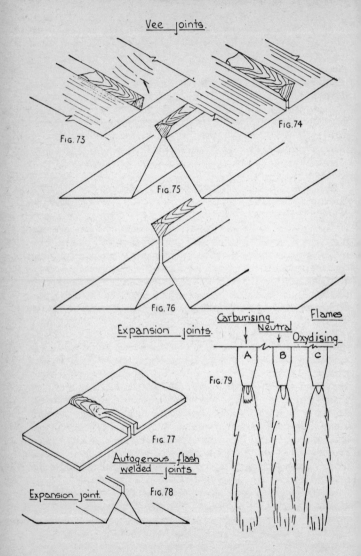

Vee joints.

Fig. 73

Fig. 74

Fig. 75

Fig. 76

Expansion joints.

Fig. 77

Expansion joint.   Fig. 78

Autogenous flash welded joints

Carburising   Flames
Neutral
Oxydising

Fig. 79

A   B   C

clear of the molten metal and at right-angles to the job. This is necessary so that the flame will protect the copper from oxidisation. No filler-rod is needed with this method, as the up-turned edges provide enough metal for the weld and also give rigidity to the joint.

In welding branch joints on copper tubes, and in many situations in copper roof-work, flash welding cannot be used and a filler-rod is essential. The method is much the same as that used both with mild steel and in lead-burning, with the precaution just mentioned that the flame should be used to prevent the formation of oxide.

Bronze Welding

Bronze welding is actually a form of brazing, and is really misnamed welding. In this process copper is jointed with the use of a filler-rod of bronze—generally phosphor bronze. The method has a great advantage in that the filler melts at a temperature much below that of copper. This spares the operator the worry of melting the copper.

As the copper is not melted, the welding depends—as in brazing and soldering—on surface fusion only, but it will be found that a bronze weld has remarkable strength. The bond between the bronze and the copper is stronger than the copper itself, if the workmanship is satisfactory.

The application of bronze welding to sheet-copper work is shown in Figs. 73–76. Butt and lapped seams have been omitted as needing no further illustration (see Figs. 33 and 34). The more common joints made on copper tubes are illustrated in Figs. 164–167. For both purposes the technique is the same. The adjacent pieces of copper are heated by means of a neutral flame, until at a touch of the filler-rod the bronze is seen to fuse with the copper. There the weld is carefully controlled by means of the flame. By flame movement, the point at which fusion is taking place can be regulated. The aim is to make a continuous weld by fusing a molten pool of bronze, first to one piece of copper, then the other. A rhythmic rotary movement is soon developed.

Bronze welding differs from brazing in that control of the bronze is possible, while brazing spelter runs widely over the surface.

Difficulty is often experienced in bronze-welding brass unions, etc., to copper tube. This arises from the composition of brass.

The alloy is a mixture of copper and zinc and perhaps a little tin. When brass is heated it gives off fumes which make the weld porous. To overcome this difficulty an oxidising flame should be used. To produce an oxidising flame, increase the supply of oxygen to a neutral flame until the cone is half its normal size; with care the weld should now be possible. If the brass happens to have a melting-point below that of the filler-rod, it might be necessary to use a brass filler-rod or a brazing spelter.

A flux is necessary in bronze welding, and reliable proprietary brands are to be recommended.

Weldable Fittings

For waste- and soil-pipe work a great deal of effort and work can be saved by the use of "weldable fittings." These are copper fittings into which the pipe ends fit against a shoulder. They are provided with a space at the mouth of each collar in which the weld is made. They can be had for most purposes.

Brazing

Brazing is a process in which brass is used for joining other metals, chiefly copper, iron, and steel. It is an ancient process, possible because any means of heat raising can be used to produce a dull red heat. This can be done in a fire, or by means of an air-gas flame, and of course by means of oxy-acetylene.

For successful brazing and a strong joint, lap-joints are essential. The process is a simple one: the metals must be heated to red heat, the surfaces sprinkled with a borax flux, and a small quantity of brass spelter applied. When the temperature of the job is high enough to melt the brass, it will be seen to spread between the pieces of metal and fuse with their surfaces.

Using a gas flame, more accurate control of the brazing is possible and a neater job will result.

CHAPTER IX

EAVES-GUTTERS AND RAIN-WATER PIPES

THE disposal of rain-water generally falls to the lot of the plumber. It can be dealt with conveniently in two sections— eaves-gutters and rain-water pipes.

Eaves-gutters have to be provided whenever a sloping roof overhangs an outside wall. The size of the gutter depends on the area of roof discharging into it.

Excepting wood, eaves-gutters are made from cast-iron, asbestos-cement, or aluminium; cast-iron being most common and probably the cheapest.

Gutters are made in many sections and some typical examples are given in Fig. 88. They vary from half round (Fig. 81) to ogee and a great variety of specially moulded patterns generally known by numbers. In size they can be had from 75 to 150 mm (3 to 6 in) wide from stock, and to almost any size when specially ordered.

The most-used gutter is probably the 177 mm (5 in) half-round beaded (Fig. 83). This gutter is found sufficient for general domestic work. The gutters are made in lengths of 2, 1·25, 1 and 0·6 m (6, 4, 3 and 2 ft) to limit the amount of cutting and wastage.

The first important consideration is the setting out of the job; of deciding on suitable falls and the position of fall pipes or down spouts. This latter factor may be to some extent predetermined by the position of existing drains, and rain-water fall pipes should, where possible, be entirely vertical.

The amount of fall in a gutter need not be great, provided the water will run off. In fact, excessive fall will tend to spoil the appearance of the building.

Having decided on the outlets and the best distribution of falls, the fixing can begin by screwing up the hangers to a chalk line. Hangers are of two main types: those screwed on to the spar feet and those screwed to a fascia-board. Both kinds prove successful, and although the first type are initially the sounder job, they are not so easily replaced. The fascia-board hangers are usually machine pressed and of galvanised malleable iron. The spar feet

Eaves gutter details.

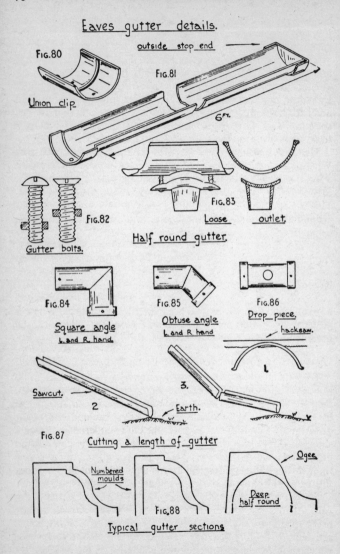

Fig. 80

Union clip

outside stop end

Fig. 81

6 ft.

Fig. 82

Gutter bolts.

Fig. 83

Loose **outlet.**

Half round gutter.

Fig. 84

Square angle
L. and R. hand.

Fig. 85

Obtuse angle
L. and R. hand

Fig. 86

Drop piece.

hacksaw.

1.

Sawcut.

2

3.

Earth.

Fig. 87 Cutting a length of gutter

Numbered moulds

Ogee

Deep half round

Fig. 88

Typical gutter sections

hangers are generally made by a blacksmith for each particular job, and of course they will vary in depth to give fall because the spar ends on to which they are fastened will be level.

The hangers should be fixed all round if scaffolding is available, and the falls checked. If scaffolding is not erected all round, greater care will be needed with the falls. The lengths of gutter should be laid, if possible, from the outlets and working to right and left. In this way the spigot ends can easily be dropped into the sockets. Working in the reverse direction, the sockets will have to be worked under the spigots of gutters already laid.

The outlet is shown in Fig. 86 to have a socket at each end. Where this is not available, a length of gutter can be fitted with a loose outlet (Fig. 83). To accommodate this, a hole must be drilled in the length of gutter equal to the diameter of the outlet. This is usually done by drilling a series of 6·5 mm ($\frac{1}{4}$ in) holes around the perimeter of the circle, and cutting out the piece enclosed, using a small cold chisel.

Where a corner—internal or external—has to be turned, special angle pieces are provided. They can be had square or to most usual angles, such as 135°, and they are made left- or right-handed.

Should two spigot ends meet at any point and no socket exists, a union-clip (Fig. 80) can be used as a double socket. Where a gutter ends, a stop-end fitting is used (Fig. 81). These can be obtained inside or outside fitting.

Cutting and Drilling Gutters

The practical method of cutting gutters can be seen illustrated in Fig. 87 in three stages. Stage 1 shows how a saw-cut is made in the sole of the gutter. The cut should just show on the inside. Stage 2 shows how the gutter is held whilst the end is brought down smartly on to a soft surface, the cut is underneath. Stage 3 shows how the gutter will snap off cleanly at the saw-cut. A good firm blow is all that is needed. It is of course preferable to saw right through if a mechanical saw is available, but seldom does this occur. For cast-iron, a coarse hack-saw blade is needed.

The gutters are secured to each other by means of gutter bolts, which also serve to tighten the joints. For 177 mm (5 in) and other medium/small gutters, a 25 × 6·5 mm (1 × $\frac{1}{4}$ in) bolt is satisfactory (Fig. 82). They have a slotted countersunk head which should fit snugly in the countersinking on the inside of the

gutter. If the bolt is fitted so that the nut is inside the gutter, it
tends to cause obstruction.

Cast-iron gutters can be easily drilled by means of an ordinary
steel twist drill and either a hand brace or a geared hand drill.
When drilling, no lubrication should be used. In marking a
gutter for drilling, a standard socket should be placed over the
spigot to ensure good centering.

The sockets and spigots of gutters should fit as closely together
as possible, so that little reliance is placed on the jointing
material. For jointing purposes putty is generally used, and the
surfaces to be jointed should be painted with good lead paint.
Red-lead paste is often recommended but seldom used, care
should be observed in the handling both of red and white lead,
which are poisonous and dangerous.

Rain-water Pipes

Rain-water pipes used to convey rain-water from the eaves-
gutters to the drains vary in size according to the quantity of
water they are likely to carry at peak periods of rainfall. They
vary from 50 to 150 mm (2 to 6 in) diameter, and are sometimes
square or rectangular in section.

For the average house 63 or 75 mm ($2\frac{1}{2}$ or 3 in) pipes are
used.

Like gutters they are made in 2 m (6 ft) lengths—from nail
hole to nail hole, on a stack. The pipe is provided with a socket
at one end (Fig. 89). Lengths of 1·25, 1 and 0·6 m (4, 3 and 2 ft)
are also available.

Rain-water pipes can be cut in exactly the same way as gutters,
by means of a saw-cut just penetrating the thickness of the metal
and a sharp downward blow to break its back.

Perhaps it should be made clear that this method is risky with
asbestos-cement pipes and gutters, and impossible with aluminium, which is not sufficiently brittle.

The pipes are fastened to the side of a building by means of
spout-nails (Fig. 94) driven into wooden plugs. The nails pass
through the holes in the ears provided on the sockets. In order
that the pipes can be regularly and completely painted, distance
pieces (Fig. 93) of cast-iron, hardwood, lead, or of short ends of
iron pipes should be used.

The wooden plugs should have little taper, and should be
driven not less than 75 mm into the brickwork joints. Unless care

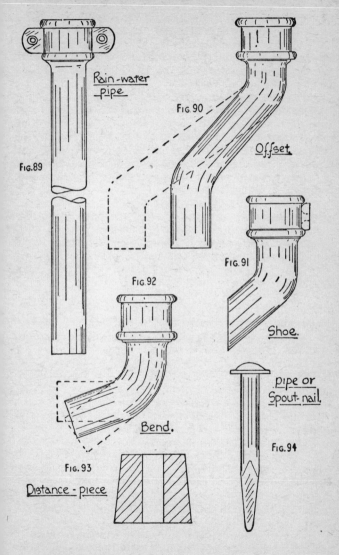

Fig. 89
Rain-water pipe

Fig. 90
Offset

Fig. 91
Shoe.

Fig. 92
Bend.

Fig. 93
Distance-piece

Fig. 94
pipe or Spout-nail.

is exercised, the plugs may lift the top few courses of brickwork, when the top lengths of rain-water pipe are being fixed. Unnecessarily thick plugs should be avoided. A little thought in setting out the job may make it unnecessary to drill the bricks, for the drilling of bricks requires much skill and patience if they are not to be fractured in the process.

It is very desirable that rain-water pipes should look plumb, and the use of a plumb-line is recommended. A word of caution here; if a long line is used, say to plumb the whole stack, it will probably be subjected to considerable wind-pressure; to counteract this a heavy weight should be used. It is probably best to plumb by the nail holes rather than attempt to estimate the centre of the pipes, or use the side of the pipe. It is, after all, the position of the nail holes which must be located.

Where a pipe must deviate from a line because of a window or some such obstruction, an offset can be used (Fig. 90). They are manufactured and stocked in steps of 38 and 75 mm ($1\frac{1}{2}$ and 3 in) so that the pipe-line can be moved from 38 to 600 or 700 mm ($1\frac{1}{2}$ to 24 or 27 in) out of line. Offsets, sometimes colloquially called swan-necks, are also used where gutter outlets are situated on overhanging eaves. If the pipe must make a considerable horizontal travel, bends are used (Fig. 92). They vary from square to very obtuse angles, to suit almost any degree of fall. Bends and offsets are not generally fitted with ears so that they are held in position by wedging with oak wedges or by caulking with tow and sometimes lead.

Rain-water pipes must—by regulation—be disconnected from the drain. They must discharge therefore, either over an open gully or into the inlet of a back-inlet gully. Where a pipe ends over a gully it should terminate in a shoe (Fig. 91). This fitting is designed to throw the water out at about the right angle. Special anti-splash shoes can also be had.

In the north of England it is general practice to discharge bath and lavatory basin wastes into a rain-water pipe. To facilitate this arrangement, a "rain-water head" is introduced. This fitting can accommodate two or three waste pipes.

The rain-water head is often used at the head of rain-water pipes, particularly where a lead-covered flat roof or a parapet gutter discharges by means of a chute (Fig. 20).

All rain-water pipes and gutters in cast-iron must be well painted inside and out with a good lead paint before fixing and need regular painting at intervals afterwards.

Rain-water pipe joints should be left open, as there is no real point in jointing them. And when open, stoppages are more easily traced and clearance effected.

CHAPTER X

COLD-WATER SUPPLY AND DISTRIBUTION

ALTHOUGH it is outside the province of the general plumber, it will be of interest to briefly consider the sources of the water we drink.

All the water we use derives initially from the oceans and is made available to us by the rain cycle. How this happens and how we utilise it is depicted in Fig. 95. There we see that water at the surface of the ocean is evaporated by the heat of the sun, to rise and form clouds of water vapour. These clouds are swept towards the land by the incoming sea breezes. Where there are hills the clouds are carried upwards into a cooler atmosphere and condensation takes place in the form of rain. The cycle is completed when the water so falling on the land drains away to streams and rivers and is carried back to the sea. Some of the rain-water, of course, will evaporate from the surface of the earth, from rivers and lakes.

Some of the rain-water soaks into the earth and is held in underground basins of impervious rock strata, from where it can be raised by means of wells. Some of this water out-pours from faults in the rock stratum and is available as a spring. The natural geological arrangement which produces the conditions in which the various types of well are found is shown in Fig. 96.

Many towns and cities use water direct from rivers and lakes. London uses the water of the Thames; Glasgow the water of Loch Katrine; Manchester the water of Thirlmere. Under such conditions of supply, very great care has to be taken that pollution is avoided and that there is sufficient purification.

Most local authorities, however, have not a large lake or river to draw upon. They find it necessary to set aside a suitable area in close proximity to the town as a catchment area, or gathering ground, from which the water can be collected and impounded in a reservoir—usually a valley having a dam thrown across it. From here the water is usually—but not invariably—piped by means of a large-sized conduit to a service reservoir on the edge

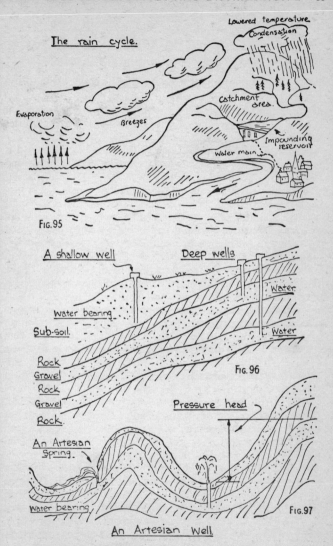

The rain cycle.

Lowered temperature
Condensation
Catchment area.
Impounding reservoir
Evaporation
Breezes
Water main

FIG. 95

A shallow well

Deep wells

Water

Water bearing

Sub-soil.

Water

Rock
Gravel
Rock
Gravel
Rock.

FIG. 96

Pressure head

An Artesian
Spring.

Water bearing

FIG. 97

An Artesian Well

of the town. At this point any needful purification or softening is generally done.

The water from moorland gathering grounds is almost always safe and soft, so soft in some cases that lime has to be added to reduce the acidity. This humic acid, originating in humus or decaying vegetation such as peat, is reckoned to be plumbo-solvent, or lead dissolving, and therefore dangerous as a possible source of lead poisoning.

On the other hand, the water from wells and borings is gener-ally hard, i.e. it contains the salts of lime, either calcium car-bonate, causing "temporary" hardness which can be removed by boiling, or calcium sulphate, causing "permanent" hardness which cannot be removed by boiling but only by a system of water softening; the base exchange method is typical.

Water softening and purification are beyond the purview of this book, and other text-books should be consulted.

Wells are of three main kinds: shallow, deep, and artesian; the latter being relatively rare in this country. The terms deep and shallow can be misleading, bearing no relationship to measured depth. It merely means that a "shallow" well only draws water from the subsoil, whilst a "deep" well must penetrate at least one impermeable rock stratum. An artesian well is necessarily a deep well in this sense. The well is bored in the bottom of a rock basin, and the water gushes out, under pressure-head supplied by water in the surrounding hills (Fig. 97).

Water Mains

Water mains are those pipes which distribute the water from the reservoir. They vary in size from huge conduits of many feet diameter to fractions of an inch.

Those of concern to the plumber are generally of cast-iron, asbestos-cement, lead, or copper.

Cast-iron pipes of 100, 75 and 50 mm (4, 3 and 2 in) diameter should be considered. They may be had in lengths varying from 2 to 4 m (6 to 12 ft) and conform to a British Standard specifi-cation. At one end there is a socket and at the other a spigot (Fig. 98). A protective coating of a bituminous kind, which is applied whilst the casting is hot and generally referred to as Dr. Angus Smith's Solution, is applied inside and out. Suitable fittings such as bends and junctions are available for use where there is a change of direction or a branch to be made.

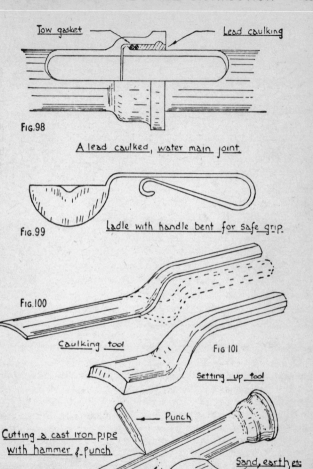

Tow gasket

Lead caulking

Fig. 98

A lead caulked, water main joint.

Fig. 99

Ladle with handle bent for safe grip.

Fig. 100

Caulking tool

Fig. 101

Setting up tool

Cutting a cast iron pipe with hammer & punch.

Fig. 102

Punch

Sand, earth, etc.

Punch marks

Cutting Cast-iron Pipes

Before attempting to cut a pipe, great care should be taken to see that accurate measurement has been taken, otherwise a faulty joint might result.

The best cut can be made by means of sawing. Whenever a sawing machine is available this has much to recommend it, but by hand it is extremely laborious and slow.

The most practicable method is by means of the pipe cutters shown in Fig. 105. For larger pipes the multi-wheel chain cutter is used, and a few minutes are all that is needed to make a cut. Care must be taken to see that the cutters are squarely fixed on the pipe, otherwise a spiral cut and possibly damage to the tool will follow.

The three-wheeled type is used on pipes not exceeding 50 mm (2 in) in diameter.

Alternatively, and in emergency for an odd cut, the method demonstrated in Fig. 102 can be used. A heavy hammer (1 kg— 2 lb) and a strong punch are needed—an old lathe-centre is just the thing. The cut is made by bedding the pipe down solidly on a heap of sand or earth, and making a succession of punch marks equidistantly on a line around the pipe. If the hammer strokes are of equal power and, say, within half an inch of each other, the pipe will generally break during a second round of blows.

Jointing Cast-iron Pipes

The socket and spigot joint is shown in Fig. 98. The method is as follows:

First the pipes must be carefully fixed, in alignment and on centre, with the end of the spigot fully abutting the shoulder of the socket. A length of hemp tow or gaskin, sufficient to pass twice around the pipe, is now inserted evenly by means of a thin caulking tool (Fig. 100), and compacted well against the back of the socket. If the pipe is lying horizontally, some means of retaining the lead filling will be needed until it has set. Traditionally, a piece of clay "puddle" is rolled and fixed in position (Fig. 103). It is now possible, however, to obtain a specially made metal ring which fits snugly around the mouth of the joint and is much less trouble (Fig. 104).

The joint is now ready for running. Molten lead is required and this can be prepared in a cast-iron metal pot, heated either

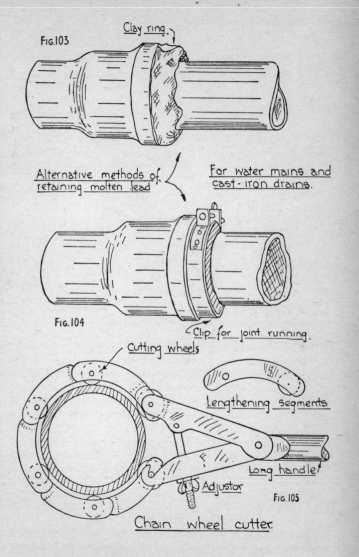

Fig.103

Clay ring.

Alternative methods of
retaining molten lead

For water mains and
cast-iron drains.

Fig.104

Clip for joint running.

Cutting wheels

Lengthening segments

Long handle

Adjustor

Fig.105

Chain wheel cutter.

on a brazier or by means of a large blow-lamp—known as a snarker, which burns paraffin oil. The lead must be very hot, so that the joint can be run before the mass of iron chills the lead to setting. Should the lead not be hot enough, the space will not be properly filled and a bad joint will result.

Molten lead being dangerous stuff and water-mains being laid in trenches, great care must be taken in its handling. For this reason, the handle of the ladle is usually, and should be, bent to provide a safe grip (Fig. 99). The ladle should be of such size that it will hold enough lead to fill the joint at one pouring.

The final caulking and finishing is done by means of a setting-up tool (Fig. 101). As water pipes are subjected to considerable pressure, the lead must be well worked in by an even distribution of drives, remembering that lead contracts on cooling and must finally form a close bond with the side of the pipe and the inside of the socket.

Because of the wetness of trenches and the great danger of molten lead flying in the presence of even the smallest amount of water, it is a wise precaution to pour a small quantity of oil—paraffin will do—into the joint immediately before the lead.

Asbestos-cement and Composition Pipes

During and since the war there has been a great increase in the use of composition pipes. They are lighter than cast-iron and can be had in correspondingly longer lengths. Like cast-iron, they are treated with a protective coating to prevent the development of algoid or vegetal growths. Such growths would by their roughness increase the friction between water and pipe, producing a loss of pressure. Specially designed joints are used on these pipes, which have spigots at both ends and loose collars with rubber rings. A special tool is used to fit the collar and rings in position.

Domestic Cold-water Supply

The bent ferrule used in tapping the main in the street is shown in Fig. 108. The water authorities use—when necessary—a special tool which is capable of drilling and tapping the main and fitting the ferrule under pressure. Close by—generally near the garden wall—a stop-cock of the screw-down or plug type is fitted (this depends on the regulations of the water authority) and a brick chamber with a cover built around it. Alternatively a drain pipe is erected above it. The tap or cock has a square or

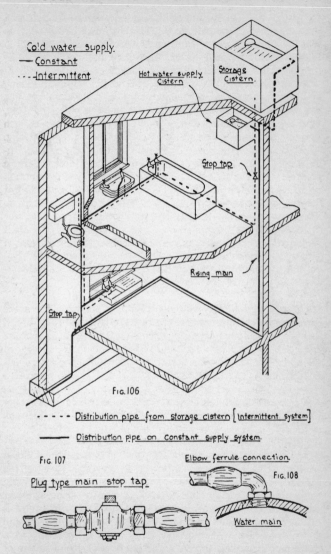

Cold water supply.
— Constant
··· Intermittent.

Hot water supply Cistern

Storage Cistern.

Stop tap

Stop tap

Rising main

Stop tap

Fig. 106

- - - - - Distribution pipe from storage cistern [intermittent system]

———— Distribution pipe on constant supply system.

Fig. 107

Plug type main stop tap

Elbow ferrule connection.

Fig. 108

Water main

rectangular top which can be reached by means of a special stop-tap key having a long handle. From the street main the service pipe is almost always of lead, but sometimes of copper of 12·5 mm ($\frac{1}{2}$ in) and 19 mm ($\frac{3}{4}$ in) diameter.

Regulations require that service pipes must be not less than 0·82 m (2$\frac{1}{2}$ ft) below normal ground level as a precaution against freezing. Particular care should be given to the depth where the pipe passes into the house. There is often a tendency to bring the pipe up into the house at an acute angle and with insufficient cover.

Stop Tap

Immediately inside the house, or any building, a consumer's stop tap should be fitted, so that all cold-water pipes in the building are under quick and easy control. This tap must be of the screw-down type—as must every other hand-operated tap on town's water (Fig. 109).

It might be mentioned here that all taps, valves, and cisterns used on water services, including hot-water services and services on meter, must be tested and stamped by the water authorities, and it is an offence to fix any fitting not so tested.

Constant and Intermittent Supply

Most water authorities provide a constant supply of water, but some few find it necessary to intermit supply for various reasons such as inadequacy of mains, capacity and demand at peak periods. This system is known as intermittent supply and for its operation, householders have to have a large storage cistern to cover their needs in periods when the supply stops.

The two arrangements are shown diagrammatically in Fig. 106. In both cases the highest point of supply is reached by the most direct route. This pipe is known as the "rising main." With constant supply, branches to the various taps are taken off the rising main where most convenient, and the rising main termin-ates at the ball tap in the hot-water supply cistern. On intermit-tent supply the storage cistern is introduced, and it is here the rising main terminates. All taps, with the exception of one for drinking water, are supplied indirectly from the storage cistern.

There is nothing to recommend the intermittent system save sheer necessity for, without great care, cisterns so often become fouled

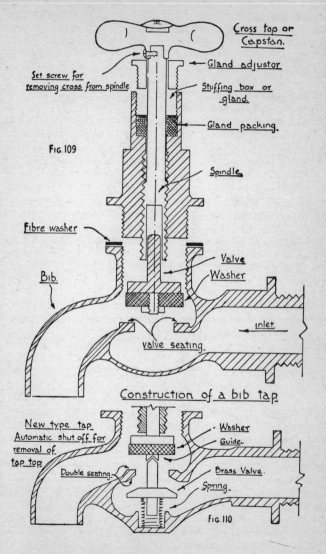

Cross top or Capstan.

Set screw for removing cross from spindle

Gland adjustor

Stuffing box or gland.

Gland packing.

Fig. 109

Spindle

Fibre washer

Valve Washer

Bib.

Inlet

valve seating.

Construction of a bib tap.

New type tap
Automatic shut off for removal of tap top

Double seating

Washer

Guide.

Brass Valve.

Spring.

Fig. 110

The storage cistern may be of cast-iron, slate, galvanised iron, or made of wood and lined with lead or copper. It should be provided with a tray or "safe" in case of leakage—to prevent damage.

Taps and Valves

Taps and valves for the cold-water supply are of three kinds: plug, screwdown, and slide.

The plug tap (Fig. 107) mentioned above is only installed in the cold-water supply, as a main stop tap. In this situation it is only rarely in operation and is quite satisfactory; it is never used elsewhere on mains, because in its operation—a 90° turn of the valve plug—the moving column of water is brought to an instantaneous halt. Now, as water is practically incompressible, and the moving water has momentum, the sudden stopping of the stream causes the energy in the water to be expended on the walls of the pipes. This effect is known as water-hammer which often produces loud noises in pipes (particularly copper) and is some-thing of a nuisance, quite apart from the stresses it sets up. The most common cause of water-hammer in pipes arises from the fitting of rubber washers on taps. Rubber washers should never be used, for cold water leather is the right material and for hot water the specially prepared red fibre washers should be fitted.

Screw-down taps or cocks are of four types: stop-cocks (Fig. 120), bib-cocks (Fig. 109), pillar-cocks (Fig. 111), and globe-cocks (formerly used on baths) and now largely replaced by pillar taps.

A stop tap provides an in-line water-way with a connection at each end for the pipe. As the water flows in *under* the seating to lift the valve, and flows out *over* the seating of the valve, care must always be taken in fixing to see that the tap is fixed the right way round. As an aid to this an arrow is usually embossed on the side, but most plumbers merely have a glance inside.

The bib tap is shown in cross-section in Fig. 109. Except for the shape of the water-way, the internal arrangements are the same as in any screw-down tap. The section shown has been exploded to show the various parts more clearly. This is the tap we find on most sinks.

The pillar tap, designed for use on lavatory basins and on some baths (Fig. 111), is similar in construction to all modern taps, and has what is called an easy-clean cover; this is shown here.

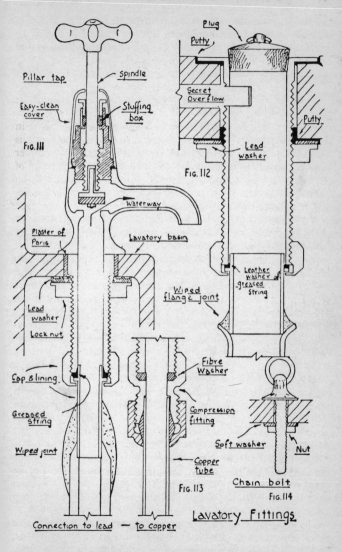

Pillar tap

Spindle

Easy-clean cover

Stuffing box

Fig. 111

Waterway

Plaster of Paris

Lavatory basin

Lead washer

Lock nut

Cap & lining

Greased string

Wiped joint

Connection to lead — to copper

Plug

Putty

Secret Overflow

Lead washer

Putty

Fig. 112

Wiped flange joint

Leather washer & greased string

Fibre Washer

Compression fitting

Copper tube

Fig. 113

Chain bolt

Fig. 114

Soft washer   Nut

Lavatory Fittings

It will be seen that the cover will have to be removed before repairs to the tap can take place.

The globe tap for baths differs only in water-way design.

The fixing of pillar and globe taps is dealt with in the chapter dealing with sanitary fittings.

Maintenance and Repairs

In essence, a screw-down tap consists of five parts (Fig. 109). They are: (A) a water-way fitted with a ground seating; (B) a top casting, made to screw into (A); (C) a spindle fitted with a crutch or capstan top, and a quick-pitched thread for the raising and lowering of the valve; (D) the valve which carries a washer and is made to fit loosely into the end of the spindle and to centre on to the seating; (E) a stuffing-box or packing gland to prevent the escape of water up the spindle.

It will be evident that the working parts will gradually be worn by use, but they do, in fact, usually remain efficient for something like twenty years—and often longer.

The stuffing-box is in need of adjustment at regular and fairly frequent intervals. To do this the easy-clean cover will have to be removed. Before this can be done, it will be necessary to remove the crutch or capstan; this component is fitted on to a square and secured by a set-screw. Having removed the set-screw, the crutch can be removed by means of a sharp blow from a piece of wood— or if a hammer is used, some kind of "softening" will be needed to prevent damage to the plating or even to the metal.

One or two turns on the gland-screw is usually enough to effect water-tightness. If the gland-screw has been successively adjusted and is well down, a new ring of packing will be needed. This consists of soft cotton string, well soaked in tallow and laid in evenly.

A leather or fibre ring—the latter for hot-water taps—is fitted between the upper and lower parts of the tap. From time to time this may need renewal, but in ordinary circumstances it will last as long as the tap.

The washer which is of leather or fibre, and should not be of rubber, will need changing from time to time, because either it has seen excessive use and is worn out, or may have become hardened. To carry out a replacement, the water must be shut off at the stop tap, and the tap should be opened to run off any water at a higher level in the pipes. The tap top is now removed

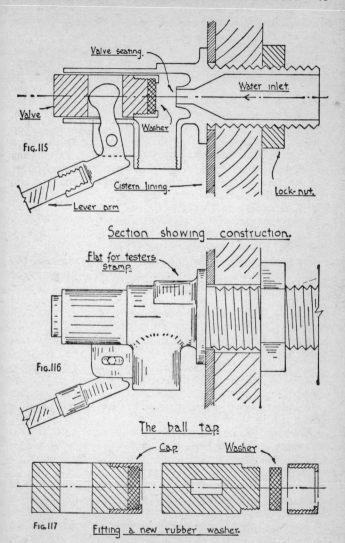

Valve seating.

Water inlet.

Valve

Washer

Fig. 115

Cistern lining.

Lock-nut.

Lever arm

Section showing construction.

Flat for testers stamp.

Fig. 116

The ball tap.

Cap

Washer

Fig. 117   Fitting a new rubber washer.

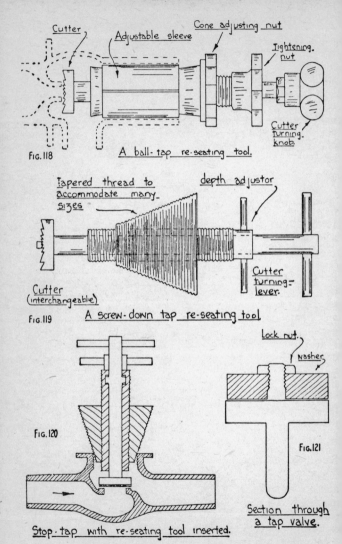

Cutter

Adjustable sleeve

Cone adjusting nut

Tightening nut

Cutter turning knob.

Fig. 118   A ball-tap re-seating tool.

Tapered thread to accommodate many sizes

depth adjustor

Cutter (interchangeable)

Cutter turning-lever.

Fig. 119   A screw-down tap re-seating tool.

Fig. 120

Lock nut.

Washer.

Fig. 121

Stop-tap with re-seating tool inserted.

Section through a tap valve.

and the valve extracted. The washer will be seen to be held in position against the valve plate by means of a small nut. When this has been removed the new washer can be fitted. Many plumbers contend that a leather-washer has a right and wrong side, if the leather is sound, this contention is unimportant.

A number of new-type taps have been introduced recently, taps which shut off automatically by means of a spring-loaded under-valve when the spindle is screwed upwards beyond a certain point. The principle is shown in Fig. 110. With this type of tap it is obviously unnecessary to shut off the supply at the stop tap.

The Ball Tap

The ball tap, cock, or valve is of the slide type. It is operated by a rising or falling water-level in a cistern. It is shown in Figs. 115 and 116, and consists of a slide valve which moves to or from a seating when moved by a float-operated lever.

The only maintenance generally needed is an occasional scraping of the moving parts, which sometimes become stiff through an accretion of substance from the water, and replacement of the rubber washer. Rubber is used in this case because of the inconsiderable force applied by the flotation of the ball and leverage, and because the rising water will apply the closure gently and so avoid water-hammer.

The dismantling of the valve is to be seen in Fig. 117. A screwdriver is held through the slot in the valve, and the cap can be removed by means of a pair of gas pliers. It should be noted that the meeting of the cap with the body of the valve is not always visible, and many amateurs attempt to stuff the washer in from the end. When fitted, the washer should present a flat surface to the seating. If the washer is too thick and is left bulging out, it is liable to quickly become distorted and to be soon damaged.

Re-seating Taps

The seats of many taps became pitted in time, from impurities and grit in the water and faults in the metal of which they are made. Re-washering is then useless, and rubber washers are often resorted to. This can never be more than a mere temporary expedient, and is to be deprecated.

In such cases, the seating can be re-ground by means of a special tool. Fig. 119 shows the one used on screw-down taps;

PIPE FIXINGS

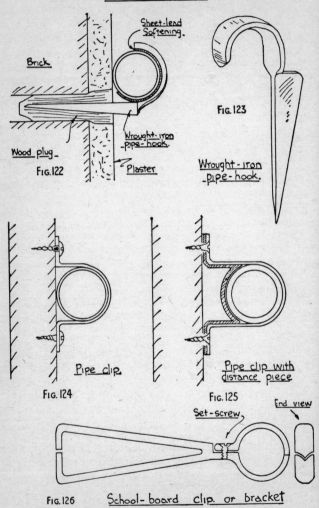

Brick

Sheet-lead Softening.

Wrought-iron pipe-hook.

Wood plug.

Fig. 122

Plaster

Fig. 123

Wrought-iron pipe-hook.

Pipe clip.

Fig. 124

Pipe clip with distance piece

Fig. 125

Set-screw

End view

Fig. 126

School-board clip. or bracket

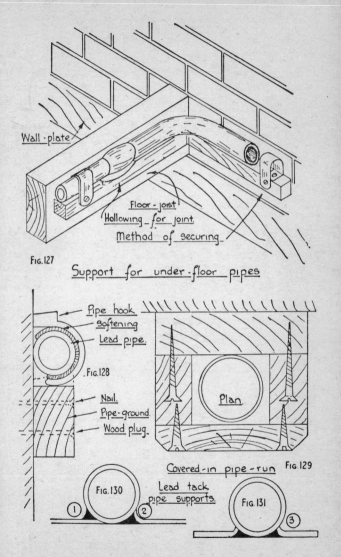

Wall-plate

Floor-joist
Hollowing for joint
Method of securing

FIG. 127

Support for under-floor pipes

Pipe hook
Softening
Lead pipe.

FIG. 128

Nail.
Pipe-ground.
Wood plug.

Plan.

FIG. 129

Covered-in pipe-run

FIG. 130   Lead tack   FIG. 131
pipe supports.

① ②   ③

Fig. 118 that for ball taps. The operation is simple and needs little explanation. The tool has a tapering thread to accommodate many tap sizes, and interchangeable cutters for each size of tap.

Taps in common use are usually of brass or white metal, but they can be had—at a price—in stainless steel. They are given for size as the pipe to which they are fixed, so that sink and lavatory taps are generally 12·7 mm ($\frac{1}{2}$ in) and bath taps 19 mm ($\frac{3}{4}$ in), whilst the domestic ball taps are 12·7 mm ($\frac{1}{2}$ in).

Fastenings

Methods of fastening and supporting lead pipes are shown in Figs. 127–131. Lead pipe requires support as shown in Fig. 127; it will be seen that the "ground"—as the support is called—has been hollowed out to receive a joint. It is convenient to bend the normal pipe clip to the shape illustrated.

Figs. 122, 123, and 128 show the use of a wrought-iron pipe hook; it will be seen that the lead pipe is protected from the sharp edges of the iron hook by means of a strip of lead used as "softening" to take up the strain. They are usually driven into a wooden plug.

Fig. 129 shows how pipes can be covered in: the pipe is clipped to a pipe board and covered as shown. In exposed places the inside space can be filled with a good insulating material such as slag-wool, spun-glass, etc.

In Figs. 130 and 131 three methods of fixing lead ears to a lead pipe are shown.

Three methods of securing copper water pipes are given in Figs. 124, 125, and 126.

The clip in Fig. 124 may be had in galvanised iron, brass, or copper, and is secured by means of screws to a wooden pipe board.

Fig. 125 is in two parts, one of which serves as a distance-piece, whilst the other holds the pipe.

The clip in Fig. 126 is designed for fixing into a wall; cement is generally used.

Flushing Cisterns

Flushing cisterns are of two main types as shown in cross-section in Figs. 132 and 133.

Both are supplied with water through a ball tap and are

SIPHONIC FLUSHING CISTERNS.

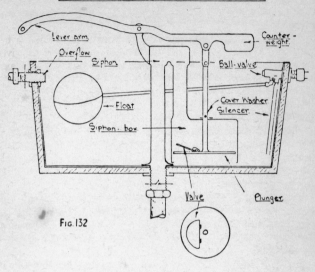

FIG. 132

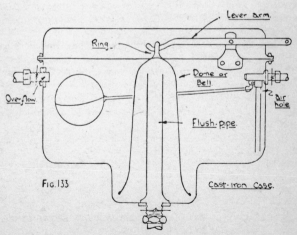

FIG. 133

fitted with an overflow which must discharge in such a way that it will be a nuisance and so receive attention. Both are siphonic in principle.

Fig. 132 shows the plunger and siphon type. Here a pull on the lever causes the plunger to throw a volume of water over the crown of the siphon to start siphonage. The plunger is fitted with a hinged valve which allows the contents of the cistern to pass through, over the siphon and down the flush-pipe.

The bell-type cistern in Fig. 133 is usually of cast-iron. In operation, the lever lifts the bell, which on falling causes a reduction of pressure in the flush-pipe by a downward rush of air.

The contents of the cistern are then forced through the bell and down the flush-pipe by atmospheric pressure on the water surface.

Frost Precautions

In this country, a severe frost always seems to take us by surprise. And our precautions against it are seldom satisfactory.

In general, pipes should be protected from draughts and they should be insulated from outside walls. A suitable method is given in Fig. 129.

Under floors and in the roof space of buildings, pipes should be wrapped with canvas backed, hair-felt strip or otherwise covered with a wrapping of spun-glass or any suitable heat-insulating material.

Cisterns exposed to cold air should be boxed in, packed around with insulating material, and fitted with a lid.

It should be noted that a pipe that has been frozen, although it may not burst immediately, will have been greatly weakened and may cause trouble later.

CHAPTER XI

SOLDERING

MANY craftsmen use solder for various purposes, but none with such diversity as the plumber. He can well be regarded as the soldering expert. Soldering is not so much a skill of habit as a mental skill. An understanding of the nature of metals and the behaviour of the various alloys of lead and tin is of prime importance. The effect of heat, the formation of oxides, and the use of fluxes are all problems needing some thought.

Soldering is a process of joining the surfaces of two pieces of less fusible metal by means of a fusible alloy. The alloy is usually one of lead and tin, and is used with copper, brass, lead, tin, iron and steel, zinc, and some other alloys.

Capillarity

Liquids will travel against gravity and great distances in any direction between two surfaces which are in close contact. And the nearer together the surfaces, the greater the distance the liquid will travel. This is known as capillary attraction or capillarity.

This principle is of great importance in soldering, and is illustrated in Fig. 137. When heat is applied to two cleaned and fluxed pieces of sheet metal and solder is applied to the edge, it will be seen to travel quickly to all other edges. The greater the pressure the quicker the penetration. It will be seen that only a small amount of solder is required to produce a strong bond between the metals.

This principle, recently applied to the jointing of copper tubes, has been applied in soldering by plumbers for a long time. The process is known as "sweating." Fig. 138 shows how a handle may be sweated on to a piece of sheet metal. The kind of results one might find are examined in Fig. 140 by taking a section through the line A–B. Stage 1 shows a sound capillary joint, using a minimum of solder. Stage 2 has a surplus of solder at the edge, this surplus serves no useful purpose and should be avoided. Stage 3 shows a failure to produce capillary attraction, and only edge soldering has taken place. The joint would be weak. Insufficient heat has been applied.

Most beginners who have difficulty with soldering either fail to appreciate this sweating principle or do not apply enough heat. The result is generally a poultice effect on the outside of the joint.

It should be realised that soldering by means of the copper bit, or soldering-iron as it is also called, is not the easiest way to learn. It is better to begin by using a small gas-flame and observing the capillary effect. Once this has been grasped, heat raising by means of the copper bit can be tackled. The bit must be of adequate size in order that it can retain enough heat to complete the job. For general purposes, a $\frac{1}{2}$ kg (1 lb) bit is useful and can be of the poker or hatchet type (Fig. 137).

Solders

The solders with which we are concerned here are known as "soft solders" and are alloys of lead and tin.

Plumber's wiping solder, consisting of 2 parts of lead and 1 part of tin, is used in joint wiping (see Chapter XII). It melts at about 230° C (440° F), depending on its exact proportions.

Plumber's fine solder is of equal parts of lead and tin, and melts at about 190° C (370 °F). This is a good general-utility solder, suitable for general brass and copper work. It is used for tinning unions.

Tinman's solder is made up of 2 parts of tin and 1 part of lead, with a melting-point of 170° C (340° F). This fine solder is used for tinplate work and for delicate soldering of all kinds.

Pewter solder has an addition of a small quantity of bismuth which lowers the melting-point below that of pewter, which is also an alloy of lead and tin.

For soldering block tin, a very fine solder is used, taking care not to apply too much heat to damage the metal.

In some districts the 1 and 1 solder is called tinman's solder, and the 2 parts tin solder is referred to as blow-pipe solder.

Lead/tin solders are the subject of British Standard No. 219, which classifies a range of soft solders. Small quantities of antimony may be used.

There is a growing tendency to speak of solders in terms of the British Standard Specification in preference to the older and often confused names mentioned earlier.

The British Standard Specification covers the following solders of particular interest in plumbing.

Grade A (35% lead, 65% tin, up to 1% antimony) has a low melting-point and is suitable for spigot joints (Fig. 154).

Grade B (48% lead, 50% tin, 2–3% antimony) for tinning brass work and fine hand soldering.

Grade C (58% lead, 40% tin, 2% antimony) for general copper-bit work.

Grade D (68% lead, 30% tin, 2% antimony) or *Grade H* (65% lead, 35% tin) for wiped joints on lead pipes, etc.

Fluxes

Many multi-purpose and specialist proprietary fluxes are on the market, and are to be recommended for the purposes for which they are intended. They are not always to hand, and the basic and traditional fluxes should be known.

For soldering lead, tallow and resin should be used; the tallow prevents oxidisation and the resin assists fusion.

For copper and brass which has been cleaned, resin is also used, especially where a non-corrosive flux is wanted. Otherwise, killed spirits of salts can be used. Killed spirits is produced by dissolving zinc in hydrochloric acid, also called spirits of salts. It has a slightly corrosive effect and must be washed away when the job is finished. This flux has an advantage in its cleansing properties when applied to hot copper and brass.

The flux for zinc is spirits of salts (unkilled), as also for galvanised iron.

For tin, Galipoli oil is recommended.

Fig. 142 shows the capillary soldering of a copper ball. It is the solder between the surfaces which is important.

Fig. 141 is a partial section of a "sweated" joint on screwed copper tube. This type of joint has been used for many years on heavy quality "screwing copper tube." The thread on the tube and the inside of the fitting is tinned, using killed spirits as a flux and fine solder. The two parts are next heated by means of a blow-lamp and tightened together. The space between the threads is now filled by sweating and where possible the solder is "floated" around the ends of the socket. By this means inspection will show any holes or flaws in the joint.

The capillary joint in Fig. 139 is the scientific development of this principle in soldering. The thread is shown to be unnecessary. The fitting is made to give a sliding fit on the copper tube, and in this case a solder ring within the socket is sufficient for filling the joint space. Before assembly, flux (supplied by the manufacturer) is applied to the surfaces. Then the joint is assembled and heat applied to the whole by means of a blow-lamp flame. The joint is satisfactory when the solder is seen to exude from all parts of the joint.

Other fittings of this type are without the patent solder insert, and the solder is fed to the joint.

The Blow-lamp

This tool is an indispensable part of a plumber's kit. Its construction and functioning should be understood, both for the sake of efficient use and for safety.

Formerly most craftsmen used petrol-burning lamps and many would still prefer to. However, petrol rationing during and after the 1939–45 war led inevitably to the wider use of paraffin-oil-burning lamps. These two types are different in that petrol vaporises much more readily than does paraffin, and makes its own operational pressure. If abused in use, it is naturally the more dangerous room-mate. The paraffin lamp, on the other hand, depends on the assistance of an air pump for its working pressure, and on a rather clumsy heating coil in its nozzle. It is a heavier lamp than the petrol type and less clean to handle.

Fig. 134 shows a diagrammatic section through a paraffin lamp. For convenience of illustration, the filler-cap, the pressure-relief valve, and the pump have been moved around. The operational principle is simple. The paraffin oil in the body of the lamp is forced by heat and the pump up the central tube and around the coil in the nozzle—where it vaporises—and out through the fine jet. Here the gas burns with a useful hot flame. The flame passes through the coil to vaporise the paraffin.

To operate the lamp, it should be nearly filled at the filler-cap, and the cap firmly screwed down to bed well on the fibre washer. The pressure-relief valve, which is actually located on the stem below the filler-cap, should now be opened. A small quantity of methylated spirit should now be poured into the well around the central tube on top. In practice, a little cotton waste, string, or even a piece of rolled paper is curled around the stem in the well, and a little of the paraffin poured on to it. After a while, when the spirit or paraffin has almost burned out, the relief valve is closed, and a light is applied to the gas when it begins to blow. Increased power is given to the flame by increasing the pressure by means of the pump.

Occasionally a stoppage may occur in the jet orifice, due to

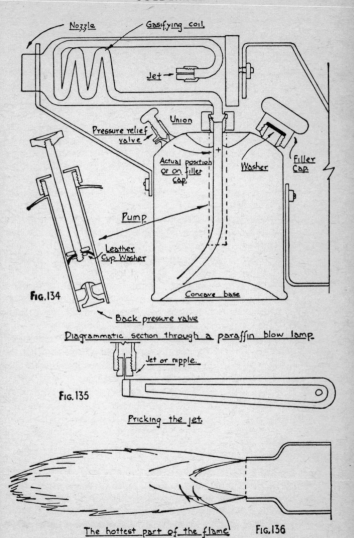

Nozzle

Gasifying coil.

Jet →

Pressure relief valve

Union

Actual position or on filler cap.

Washer

Filler Cap

Pump

Leather Cup Washer

Concave base

FIG. 134

Back pressure valve

Diagrammatic section through a paraffin blow lamp

Jet or nipple.

FIG. 135

Pricking the jet.

The hottest part of the flame   FIG. 136

dirt in the fuel. This can be cleared by using a pricker—a tool supplied with the lamp. This operation is shown in Fig. 135. If pricking fails to clear the jet, the nozzle will need to be removed and the jet unscrewed by means of a small box-spanner.

Fig. 136 indicates the hottest part of the flame, as a guide to the distance the nozzle should be held from the object to be heated.

Aluminium Soldering

Aluminium soldering has generally been avoided by plumbers, and, indeed, it has often been claimed that it cannot be done. Once more some little knowledge of science helps in discovering the reasons for this difficulty.

Aluminium readily oxidises when heated; that is, a skin of aluminium oxide is formed by a combination of oxygen from the air and the metal. In the case of every other common metal we can use a flux which will destroy or dissolve the oxide. But in the case of aluminium, it is insoluble and no suitable flux exists.

A special solder is needed and can be bought from the solder makers.

These solders have a fairly low melting-point and a long plastic period—like plumber's solder.

The soldering technique is as follows. Melt a pool of the solder on to the parts to be soldered and keep the solder in a molten condition, either by using a lamp (preferable) or a copper bit. Whilst the solder is molten, scrape the surface underneath it by means of a sharp tool. The solder will exclude the air and oxygen and will immediately adhere wherever a scratch or scrape is made. By continuously applying heat and scraping beneath the molten solder, it is not difficult to get the whole surface "tinned." Once the "tinning" has been managed, soldering can proceed as with any other metal.

Solder Paint

An interesting innovation has been the recent development of "Fryolux" solder paint. It is claimed that the time-honoured methods of tinning and soft soldering are slow and costly. And this material is an attempt to economise in time and material. It is admirably suitable for all kinds of "sweating" and the sound principle of capillary soldering. '

"Fryolux" is a uniform mixture of powdered solder and liquid

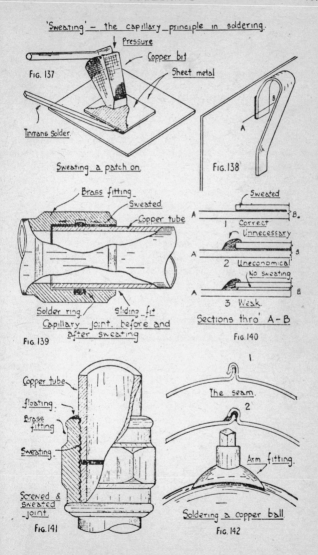

'Sweating' — the capillary principle in soldering.

FIG. 137

↓ Pressure

Copper bit

Sheet metal

Tinmans Solder.

FIG. 138

B

A

Sweating a patch on.

Brass fitting

Sweated

Copper tube

Solder ring.   Sliding fit

Capillary Joint. before and
after sweating

FIG. 139

A ————————————— B
1 Correct
Unnecessary
A ————————————— B
2 Uneconomical
No sweating
A ————————————— B
3 Weak

Sections thro' A – B

FIG. 140

Sweated

Copper tube

floating

Brass
fitting

Sweating

Screwed &
sweated
joint.

FIG. 141

1

The seam.

2

Arm fitting.

Soldering a copper ball.

FIG. 142

flux, it has a creamy consistency and is applied cold to the work, like paint.

For tinning, the article is brushed or sprayed with the paint and heated until the solder melts and forms an adherent coating. For assembling such articles as capillary joints, both surfaces are "painted" before assembly, assembled, and heated.

Overheating should of course be avoided.

Hard Soldering

This process is worth mention, and is useful where soft soldering has insufficient strength or would not withstand a high enough temperature. A number of hard solders are available, but generally silver solder is used. The operation is comparable with brazing, but of course the melting-point is lower and bright-red heat is not necessary in silver soldering.

Silver solders generally consist of mixtures of copper and silver (10–80% of silver). An alloy without silver containing 92% of copper and 8% phosphorus and melting at 705° C, is also used.

The process is very suitable for jointing copper to copper, copper to brass, and brass to brass. A flux is needed, and borax or a borax mixture is used. Because of the tendency of a dry flux to be blown away by the flame, it is well to make up a paste or add the flux to hot water. The solder can be bought in sheet strip or rod form. The sheet can be cut up into strips of the desired width, and for tube jointing widths of $6\frac{1}{2}$ mm ($\frac{1}{4}$ in) are most generally suitable.

CHAPTER XII

METHODS OF JOINTING LEAD PIPES

The Wiped Joint

JOINT wiping is a technique of great antiquity—lead pipes have been joined together by this means for at least 2,000 years. It was known from early times that an alloy of lead and tin has a melting-point rather lower than either of its constituents. It is this fortuitous circumstance that enables the craftsman to shape a joint without melting the pipe. By methods of trial and error, it was found just how much of each metal was needed to make an alloy having the best plastic properties, and which would remain molten for a time sufficient for the wiping of the joint.

The Solder

Plumber's solder is the name usually given to the alloy described above. It is also known as coarse solder and as "metal." The proportions used are approximately: 2 parts of lead to 1 part of tin and it is manufactured in bars of about ½ kg (1 lb) weight. Reliable brands should always be used as very small quantities of impurities render the joint wiping difficult and may result in faulty joints. All solder should conform to British Standard Specifications.

Blow-lamp and Pot and Ladle Methods

There has been a considerable change in practice over the past fifty years or so. Formerly, before the days of the blow-lamp, the only heating agency sufficient for the melting of solder was the fire. Consequently a technique was developed wherein the solder was melted in a cast-iron pot and applied to the lead pipe by means of a ladle. When the pipe was hot enough, surplus solder was wiped away, and the wiping of the joint was completed.

Occasionally this method must still be used in situations where the use of a blow-lamp might be a fire danger.

In general though, joint wiping today is an adaptation of the older method to the handiness of the blow-lamp. This method will be described in detail below. It must be emphasised here that joint wiping is essentially a skill, and that although considerable understanding and knowledge is needed, the manual dexterity and speed involved is such that success can only follow a fair amount of practice. Small and medium-size joints will be considered.

Some of the tools used are illustrated, and will be referred to as they are needed in the process.

Preparation

When beginning to prepare a joint, care should be taken to see that the ends are straight and square and cleanly cut. There is an economy of time and effort if this point is borne in mind when a piece is cut from a roll of pipe. And it is well to remember that *twice* measuring and *once* cutting is easier than *once* measuring and *twice* cutting. The preparation of an underhand joint is shown in Figs. 143–148 in stages.

An underhand joint is any joint which is within 45° of the horizontal. Inclined beyond this, it is referred to as an upright joint.

The section in Fig. 148 shows how one pipe is made to fit neatly within the other in such a way that there is no reduction of the pipe thickness and little increase in outside diameter at the joint. The bore of the pipe must not be in any way restricted.

One end (Fig. 145) is known as the socket and the other (Fig. 147) the spigot.

Fig. 143A shows how the socket is formed by the use of a tanpin, a hardwood cone which will open the end of the pipe without damage and will wear well. The end of the pipe should be opened out until the inside diameter of the end is equal to the outside diameter of the pipe; in this way a close-fitting joint is assured.

A cabinet rasp is now used to cut away the surplus lead on the outside of the socket. The rasp should be held parallel with the pipe, and made to cut with firm clean strokes. As it is usually necessary to hold the pipe end in a steady position with one hand, some degree of skill has to be developed in one-handed rasping.

The spigot end should now be shaped to fit neatly into the socket. Fig. 146 indicates the angle at which the rasp should incline, this will be seen to be the angle of taper of the tan-pin.

This point calls for special regard so that a good fit will be achieved throughout the length of the socket. The lead should be taken off at this angle until a small thickness of wall remains at the end. A feather edge should be avoided because it will result in an irregular end.

The fit of the ends should now be tried and any necessary adjustments made.

Although the grey film of oxide found on lead has to be removed before soldering can take place, it often happens that some solder will adhere. To prevent this, the pipe is coated for a few inches on each side of the joint with plumber's black (also known as "soil," "smudge," and "tarnish"). This black composition consists of lamp-black in glue-size or other binding medium, and is applied by means of a brush.

If the pipe is at all greasy, it is helpful to chalk on the surface before blacking; this action counters the grease and ensures a good coating.

In order to ensure a sound joint, that part of the pipe comprising the joint has to be shaved clean of lead oxide. The shavehook (Fig. 145) is specially designed for this purpose. It should be stressed that this is a *cutting* and not a *scraping* tool, and a good edge should be maintained. The angle of inclination needed to produce a good cut will readily be found by practice.

Two schools of thought exist about shaving, each preferring its own method. The one insists that the whole of the pipe end should be tarnished and the shaving carried out to a line around the pipe. The other protests that the pipe will be weakened at the line at which the shaving starts because of the tendency to dig in at the beginning of a stroke. It is therefore claimed that the shaving should be done in an irregular fashion with the black carried out to the same line. Everything considered, this seems to be a mere theoretical quibble. The method first described has an advantage in that there is the double tarnishing provided by the oxide coating and the applied black.

The length of a joint is arbitrary, but certain most suitable lengths have been arrived at over the years. They are given at the end of the chapter.

Securing the Joint for Wiping

The ends of the pipe to be jointed will need to be securely and rigidly held together while wiping takes place. Were this precau-

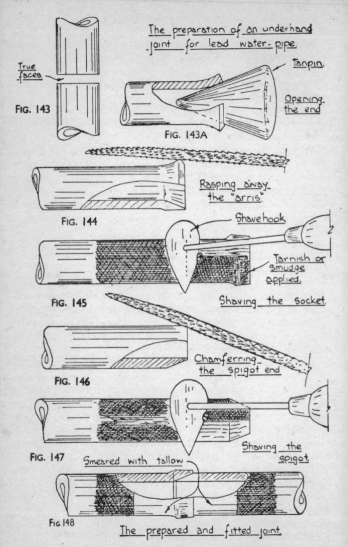

The preparation of an underhand-joint for lead water-pipe.

True faces

FIG. 143

Tanpin.

Opening the end

FIG. 143A

Rasping away the "arris"

FIG. 144

Shavehook

Tarnish or smudge applied.

FIG. 145

Shaving the socket.

Chamferring the spigot end

FIG. 146

FIG. 147

Shaving the spigot.

Smeared with tallow

FIG. 148

The prepared and fitted joint.

tion neglected, the pressure used in modelling the joint, or any vibration, would tend to crack it.

Patent pipe clamps of various types are available for this purpose, and considering their handiness, they are worth carrying about. However, with a little ingenuity they can be dispensed with if necessary, and the same effect can be obtained by means of a couple of spikes and some string, or in many cases by weighting the pipes.

The Lamp Method of Joint Wiping

With the invention of the blow-lamp, joint wiping has been much simplified. At the same time a great deal of slipshod workmanship has been allowed to develop. The method of wiping the joint is exactly the same with pot and ladle as with the lamp. The difference lies in the technique of heat raising. In whichever method is used, it is necessary to raise the temperature of the lead to the melting-point of the solder so that fusion will take place at the surface, and enough heat will need to be put into the pipe to maintain the solder in a plastic state long enough to allow of wiping.

Wiping the Joint

Immediately the ends of the pipe have been shaved, and before they are brought together, the inside of the socket should be lightly pared out by means of a pocket knife so that a clean surface is presented.

This done, all shaved surfaces should be lightly smeared with tallow to prevent oxidation or carburation by the lamp flame. The joint should now be assembled and firmly secured.

The fundamental strength of the joint depends on the goodness of fit of the ends of the pipe to each other, and the filling of the space between the joint surfaces with solder by capillary attraction. If this part of the jointing were done satisfactorily and with a good taper, it could well be argued that the subsequent wiping would be quite superfluous. The wiping does, however, give additional strength and a more sure watertightness.

The flame of the blow-lamp is now applied to the lead pipe on each side of the joint, by traversing backwards and forwards. Periodically the stick of solder is lightly rubbed on to the shaved surface until, when the pipe is hot enough, it will be seen to melt on to the pipe; "tinning" the surface by spreading. The whole

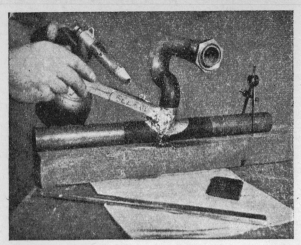

Applying the Solder to a Branch Joint.

One-handed Method of wiping an Underhand Joint.

Note.—Butane gas blow-lamp with replaceable gas container.

PLATE 3

of the surface of the joint should now be "tinned" by working the stick of solder all over the shaved portion. Particular care should be taken that sufficient solder is applied at the mouth of the socket to allow the capillary filling. When the tinning has been completed, enough solder is melted on to the pipe to allow of a satisfactory joint. This ability to estimate is acquired by a little practice.

Now the pipe should be evenly heated on each side of the joint for a distance of about 150 mm, until the solder on the joint runs off on to a moleskin wiping cloth held below in one hand. The lamp should now be laid aside, and the wiping executed quickly but deliberately. It is very important that the heat should be applied to the pipe on each side and not directly on to the joint itself. Only in this way can we be reasonably sure that there is enough heat in the whole mass of metal to maintain a plastic state in the solder for long enough. Most beginners fail through heating the solder itself, which runs off, whilst the pipe remains fairly cool and will quickly absorb the heat in the solder during wiping. Expert craftsmen, on the other hand, seem to have all the time in the world for wiping, merely because they have put the heat into the pipe, letting the solder take care of itself.

When the solder has been melted on to the wiping cloth of six or eight thicknesses of moleskin cloth, one of two methods of shaping and wiping can be used. In the first, one hand only is used (Plate 3), the fingers being so placed that the cloth is cambered to the desired shape of joint. Wiping generally begins at the front top, proceeds down the back, then from underneath and up the front. A thin piece of rag is then used to wipe off any surplus solder to the edge of the joint. Care is taken that each separate wiping overlaps the last.

With the second method two hands are used and better symmetry is achieved. Using this technique, the wiping is usually done in three movements. From the back, under, and up the front; from the front, over, and down the back; and finally under and up the front with a clean sideways wipe off.

Many beginners have the misfortune to produce "weeping" joints. The effect is usually a fine spray or beads of moisture oozing out. This is almost invariably due to reheating the solder continuously while the joint is in process of wiping. Once the solder has been melted off on to the wiping cloth, there should be no further need to use the blow-lamp, and this practice merely

LEAD PIPE JOINTS

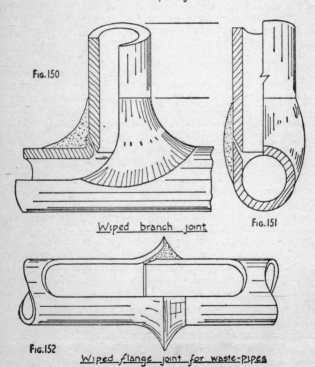

Fig. 149

Lead water pipe

Underhand wiped joint

Fig. 150

Wiped branch joint

Fig. 151

Fig. 152

Wiped flange joint for waste-pipes

indicates a lack of skill or faulty training. If solder is reheated when in a plastic state, there is a tendency for the eutectic alloy (the lowest melting-point mixture of the metals) to run off from the crystals of lead which solidify first. When this happens, the joint is left porous. Incidentally, the fine metal dripping off the bottom of the joint, and generally referred to as tin, is actually the eutectic alloy—which has the lower freezing or setting temperature—leaving the solidifying lead.

Upright Joints

Joints in the vertical position are rather more difficult to manage than underhand ones, firstly, because the solder, when melted, tends to fall away, and, secondly, because the eutectic alloy tends to leave the coarse lead crystals at the top of the joint. In the case of the underhand joint, the finer metal is wiped back on to the top of the joint, from where it will once more permeate the coarser metal.

In the case of the upright joint, more careful heating is needed, so that the solder itself is not over-heated and takes up heat only from the pipe. If reheating is required—and in many situations it is unavoidable—the heat must be applied to the pipe.

Increased speed and dexterity are needed with this type of joint, but in all other respects the method is exactly the same as that used with underhand joints. Whenever the joint can be arranged in a horizontal position, the upright joint should be avoided.

Branch Joints

The branch joint is probably the easiest of wiped joints to manage, but is often faultily prepared.

A hole is first cut into the main pipe with the aid of a branching gouge. Care should be taken that the back wall of the pipe is not damaged by the point, otherwise a potential burst exists.

Having cut a hole about the size of the bore of the branch pipe, the hole is flared out by means of a hammer and bent-pin or bending iron. This bent bolt must be inserted, so that the bent end is well round the inside lip of the hole, this precaution will avoid the formation of a ridge inside the pipe.

The hole is now opened out to form a socket for the spigot of the branch pipe. Again, a good fit is important, lest solder should be pressed through to form an obstruction (Fig. 150).

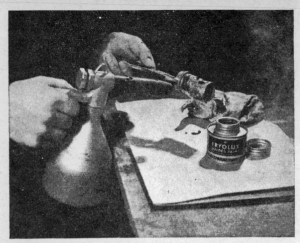

Union Tinning with Solder Paste and Blow-lamp.

Union Tinning by means of Copper Bit. (Poker type shown.)

PLATE 4

The main pipe should now be shaved around the hole, as shown in Fig. 150. To do this with care, a scribing plate and a pair of compasses are used, but with practice these can be dispensed with and a satisfactory free-hand line can be drawn.

The method of heating described for underhand joints applies equally to branch joints, and, as before, once the solder is at the correct temperature the lamp should be put aside.

In wiping the joint, the solder is first modelled quickly to a rough shape and the solder around the branch pipe is wiped to a clean edge. The wiping can now be completed in two wipes: from back to front over each shoulder in turn, care being taken to see that the wipings overlap at the back and in front. In the wipings, further care should be taken that the outside edges of the joint on the main pipe are wiped to a clean edge. There is a finger skill involved in the wiping of a branch joint, whereby the shape of the cloth gradually changes to make the joint convex on the sides but concave on the shoulders.

All joints should be left undisturbed for a few minutes after wiping, so as to avoid any risk of breaking them.

Wiped Joints to Brass Union Tailpieces

Before a brass union—or cap and lining, as it is otherwise called—can be wiped on to a lead pipe, its surface must be prepared by "tinning." In other words, a coating of solder must be fused into the surface metal. As the heat used in wiping the joint is not great enough to do this, a special operation is needed.

Tinning a Union

First, the dull coating of chemical impurities must be removed by filing. For this operation the end of the tailpiece to be tinned is held on to the corner of a block of wood, and the predetermined length to be tinned is scored by the corner of a medium-cut file. The union is slightly rotated at each stroke of the file, so that gradually every part of the surface is cleaned. Particular care must be taken that a strip of uncleaned brass is not left between successive strokes of the file, otherwise the solder may not adhere and a passage will be provided for leakage.

The brass should now be tarnished to a line which will form the edge of the joint to be wiped. When the tarnish has dried, the filed portion is lightly smeared with tallow and dusted with powdered resin.

The Staern Joint.

Note.—The special tools on the left.

PLATE 5

Powdered resin is the most suitable flux for this purpose, although a number of very good proprietary fluxes are available in paste form, obviating the need for using a coating of tallow. Liquid fluxes of the killed spirits of salts type should be avoided, because any traces left will react with the brass and cause corrosion.

Fine solder (Grade B, British Standard Specification No. 219) is used for tinning; this description is applied to the general-utility solder used by the plumber. It is made up of equal parts of lead and tin.

Heat is applied by means of a copper bit, preferably of the hatchet type and weighing not less than $\frac{1}{2}$ kg. Smaller copper bits should not be used, because they will not hold enough heat for the completion of the tinning.

A little solder is picked up on the tinned face of the bit and held against the brass. When the brass has been heated sufficiently, the solder will be seen to spread along the surface. The bit should now be slowly drawn along the filing from one end to the other whilst the union is slowly rotated at the same time. This rotation should be so contrived that the surplus solder is spreading "downhill" in front of the traversing bit.

Careful inspection should be made to ensure that every part is properly tinned.

An alternative method of tinning is shown at the top of Plate 4.

Securing the Union for Wiping

A piece of wood dowelling for small unions, or three or four pieces of stick for larger unions, can be used in securing the union to the lead pipe. These are passed through the union so as to wedge themselves inside the pipe. The union nut should be held in position by means of a small wood screw inserted between the thread of the nut and the flange of the lining.

With the above procedure, clamps are not needed, but if clamps are available, they can be attached to a wooden plug pressed into the end of the union, and left 150 mm or so on the outside.

The table on page 125 indicates suitable lengths for joints on various pipe sizes, with the approximate amounts of solder used.

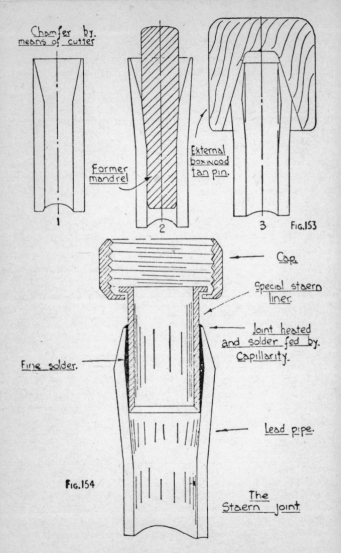

Chamfer by means of cutter

Former mandrel

External boxwood tan pin.

1   2   3   Fig. 153

Cap.

Special staern liner.

Joint heated and solder fed by Capillarity.

Fine solder.

Lead pipe.

Fig. 154

The Staern joint.

Internal Diameter of Pipe		Length of Joint		Weight of Solder used	
mm	in	mm	in	kg	lb
12·7	$\frac{1}{2}$	70	$2\frac{3}{4}$	0·225	$\frac{1}{2}$
19	$\frac{3}{4}$	70	$2\frac{3}{4}$	0·34	$\frac{3}{4}$
25·4	1	76	3	0·45	1
32	$1\frac{1}{4}$	76	3	0·57	$1\frac{1}{4}$
38	$1\frac{1}{2}$	83	$3\frac{1}{4}$	0·68	$1\frac{1}{2}$
51	2	83	$3\frac{1}{4}$	0·9	2
63	$2\frac{1}{2}$	83	$3\frac{1}{4}$	1·02	$2\frac{1}{4}$
76	3	89	$3\frac{1}{2}$	1·13	$2\frac{1}{2}$
89	$3\frac{1}{2}$	89	$3\frac{1}{2}$	1·25	$2\frac{3}{4}$
102	4	89	$3\frac{1}{2}$	1·36	3

The foregoing discussion has been concerned with joints of the sizes most generally met with on water-supply and on waste-pipes. That is, up to 25 mm (1 in) diameter lead water pipes and up to 38 mm ($1\frac{1}{2}$ in) lead waste pipes.

Because of the difficulty of raising sufficient heat for the wiping of large-sized joints, reheating is unavoidable, and the joint must be wiped in stages. However, the problems involved in large-joint wiping are many and beyond the scope of this book, and it is most unlikely that the beginner would be called on for such work.

The Staern Joint

Fig. 154 shows the Staern joint in which the capillary principle is applied to the jointing of lead pipes, in this case to a brass union. Special tools are used as seen in Fig. 153, and heat and solder are applied as shown in Plate 5. There is much to recommend this joint for its economy of material and sound principle.

CHAPTER XIII

COPPER-TUBE JOINTING

THE copper tube or pipe used by plumbers can be divided into three classes—"heavy screwing," medium gauge, and light gauge. Heavy-gauge tube ranges from 4 or 6 to 12 S.W.G. The medium gauge is usually 14 S.W.G. (about $6 \cdot 5$ mm—$\frac{1}{16}$ in—thick), and light gauge already in use varies from 17 to 19 S.W.G., but recently a British Standard Specification has been established and has gained general acceptance.

Heavy-gauge copper tube is to be found in use, but is seldom installed today except for special work. The pipe is treated in exactly the same way as mild-steel pipe, the thread used being the standard gas thread. The joints are generally made with red or white lead, or another suitable jointing paste, or may be sweated (see Chapter XI). For greater security a few strands of hemp or cotton fibre may be wound around the threads before screwing into the fitting. This heavy pipe is easily bent, but suitable brass or gun-metal fittings are available for all purposes.

Medium-gauge copper tube (Fig. 156) is jointed as described in Chapter XI by threading and sweating. The thread used, being somewhat finer than the gas thread, penetrates less deeply into the wall of the tube and the assembled joint has considerable strength.

Light-gauge Tube Joints

The compression joints provide the easiest method of jointing light-gauge tubes, and in many ways they are highly satisfactory. They have a useful advantage, in that they can be so easily taken apart, and repairs are quickly executed. They are generally clumsy in appearance and rather expensive. They are, however, more universally used than any other kind of joint.

There are two main types—those in which the pipe ends are opened or flared by means of a special tool, and those with plain pipe ends.

The first type is shown in Figs. 161 and 162. Fig. 161 is an "exploded" view of a compression fitting, in which the pipe end is flared to fit exactly over the body of the fitting. The flared end

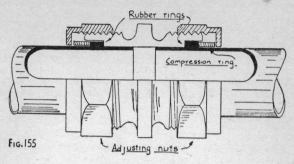

Rubber rings

Compression ring.

FIG.155

Adjusting nuts

Gland type of compression joint

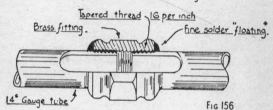

Tapered thread 16 per inch

Brass fitting.

fine solder "floating."

14ᵉ Gauge tube

FIG 156

Screwed & Sweated joint for medium gauge tube.

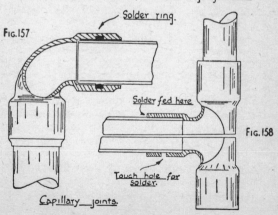

Solder ring.

FIG.157

Solder fed here

FIG.158

Touch hole for solder.

Capillary joints.

is tightened on to the body and held in place by means of a collar drawn up by a hexagonal nut. This is an expensive but highly efficient joint. Fig. 162 illustrates another pattern, utilising an internal brass cone. The ends of the pipe are here again flared and clamped on to the cone by means of the hexagonal nut and the turned surfaces in both halves of the fitting. For flaring the pipe ends, a special steel drift is used (Fig. 163). These tools are provided by the makers and must be identical in taper with the inside of the fitting.

This type of fitting is not so much used today as it formerly was, but every plumber should consider them for purposes of maintenance; there are very many in existence.

The "plain end" fitting has largely superseded the type discussed above, mainly because of the simplicity of installation, but partly for reasons of cost.

Typical examples are pictured in Figs. 159 and 160. The principle is a simple one; in short, a cone, barrel, ring, or a piece of metal in some other annular form is wedged between the pipe and the body of the fitting. It is important that this "wedge" should fit exactly to the taper of the fitting and the side of the tube, and should have as large a bearing surface as possible. Fig. 159 shows such a joint assembled, and exploded to show the parts. Fig. 160 shows alternative types, using (1) a copper "cone," (2) a copper "barrel," and (3) a fluted brass ring.

To assemble these joints, any burrs should be carefully removed from both the inside and outside of the tube. The tube end will be found to slide neatly into the fitting until it stops against the inner shoulder. At every stage care must be taken that the tube is kept up against the shoulder, or the joint will probably leak. Having tried the tube end in the fitting, it should be removed, and the cone or barrel pressed on and into place. The joint is now reassembled and the cone squeezed in by means of the hexagonal tightening nut. To tighten fittings effectively, two spanners should be used on opposite nuts. It will be found in practice that considerable force is needed to make an effective joint.

Copper tubes can be cut to length by means of a fine hack-saw blade. If large-toothed blades are used, the teeth tend to break when cutting thin-walled tubes. Handy rotary cutters can also be used. They are similar in type to the tool shown in Fig. 212. There is, however, only one cutting wheel and two rollers. Where

COMPRESSION JOINTS FOR COPPER TUBES

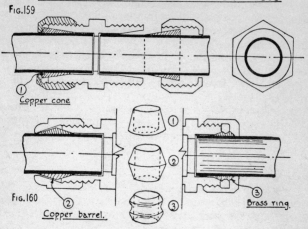

FIG.159

① Copper cone

FIG.160

② Copper barrel.

① ② ③

③ Brass ring.

Typical compression joints with plain tube ends

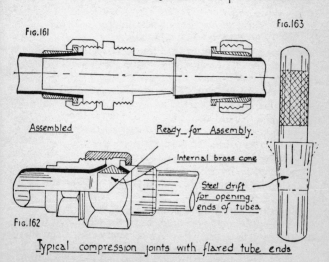

FIG.161

FIG.163

Assembled

Ready for Assembly.

Internal brass cone

Steel drift for opening ends of tubes.

FIG.162

Typical compression joints with flared tube ends

a great deal of cutting must be done by means of the saw, a special vice should be used. This tool clamps over a section of the tube, and the saw blade operates through a slot. This arrangement ensures a square cut, little burring of the ends, and fewer broken blades.

In some circumstances, it is desirable and even necessary to provide a means of taking up expansion and contraction. Waste pipes taking an alternation of hot and cold water are a case in point. To meet this need, a gland-type compression joint can be used. Such a joint has been partially sectioned in Fig. 155. In this joint the copper or brass ring is replaced by a rubber one. This allows movement either way. If after a while slight leakage occurs, the gland can be tightened up.

Capillary Joints

The capillary principle in soldering has been discussed in Chapter XI. The application of this principle in the jointing of copper tubes has been a notable forward move. These joints, devoid of hexagonal nuts, are very much neater in appearance than compression joints. If made in material of strength equivalent to the copper tube, the fitting need be of no greater thickness. Typical joints are shown in Figs. 157 and 158. The first example has a pre-formed solder ring which when heated supplies the joint space with solder. Fig. 158 illustrates a tee-fitting; for convenience it has been halved and sectioned to show two methods of introducing the solder to the joint space. The upper half shows how the solder is applied at the end of the socket. The lower half has a "touch hole" provided.

The success of a capillary joint depends on these important factors.

(1) Both surfaces must be thoroughly cleaned by means of steel wool or other gentle abrasive.

(2) A non-corrosive flux must be applied to all parts.

(3) The pipe should be a sliding fit in the socket of the fitting.

(4) The end of the pipe must fit hard up to the shoulder of the socket.

(5) Special solder or tinman's solder should be used.

(6) The joint should be thoroughly and evenly heated. Difficult with large fittings.

(7) The solder should appear as a ring all round the end of the socket.

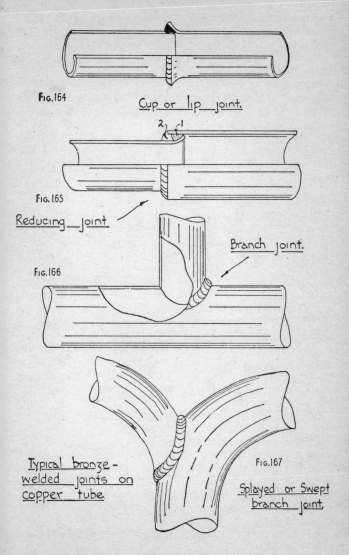

Fig. 164

Cup or lip joint.

Fig. 165

Reducing joint

Fig. 166

Branch joint.

Typical bronze-welded joints on copper tube

Fig. 167

Splayed or swept branch joint.

Capillary joints are commonly used on pipes up to 50 mm (2 in) diameter in this country, but in America they have been used on pipes of 100 mm (4 in) and greater diameter.

Bronze-welded Joints

This means of jointing copper pipes has been discussed in Chapter VIII. The more usual joints are illustrated in Figs. 164–167.

CHAPTER XIV

SOLID-DRAWN LEAD-PIPE BENDING

LEAD-PIPE bending is a relatively modern plumbing skill. Formerly bends on large-bore pipes were made by bossing two bent half-pipes and seaming them together by soldering. Increased knowledge of the nature of the metal, and the spread and sharing of experience and ideas in classes, have enabled us to perform operations of skill undreamed of by most of our predecessors.

A skilled craftsman today would have no difficulty in making a short-radius, square bend on a 100 mm (4 in) lead pipe in half an hour. And a saw cut across the bend would show it to be regular in thickness.

Lead-pipe bending is generally involved in waste- and soil-pipe installations, and the pipes range from 32 to 100 mm (1¼ to 4 in) in diameter.

Many alternative methods of pipe bending have been tried with varying success—from loading with sand and even water and by means of spiral springs. It has also been suggested that if bent quickly enough it wouldn't have time to kink! Loading is unsatisfactory. The use of springs is not to be recommended, because the method is unsound in principle, as will be shown. The bending takes place by merely stretching the back of the bend.

Most waste-pipe bends can be made by judicious use of a dresser and a little heat—unless sharp bends are required. The centre of the bend and its length should be marked on the pipe, using chalk. The part to be bent should then be dressed until it is oval in section. Now the pipe should be gently heated by means of a gas flame, particularly at the sides, i.e., the flattened part. The pipe is now taken in the hands and pulled to the shape required. The bending will swell out the flattened sides. If the bend is fairly sharp, this operation may have to be repeated twice or more. In this case the lead on the sides should be driven towards the back when dressing. If the pipe should be reduced in size at the bend, a hardwood bobbin can be driven through (Fig. 174) to restore the bore.

In bending 50 mm (2 in) diameter pipes, bobbins will almost

Solid-drawn lead-pipe bending tools.

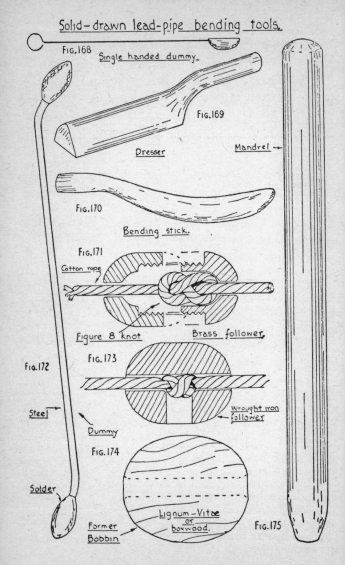

FIG.168

Single handed dummy.

FIG.169

Dresser

Mandrel →

FIG.170

Bending stick.

FIG.171

Cotton rope

Figure 8 knot   Brass follower.

FIG.173

FIG.172

Wrought iron follower

Steel →

Dummy

FIG.174

Solder

Lignum-Vitæ or boxwood.

Former Bobbin

FIG.175

certainly be needed and, should the bend be of long radius, a set of wooden followers will be required with which to drive the bobbin. Followers should be nearly spherical and only slightly less than the bobbin. They should be driven by means of a stick and hammer.

As the diameter of the pipe to be bent increases, its walls become relatively thinner; in other words, the surface area increases at a far greater rate than does the thickness of the metal. Consequently, much greater care is needed, and more attention must be given to moving surplus lead from the inside, to the stretching and thinning outside of the bend.

The problem is illustrated by Figs. 177 and 178. In Fig. 177 the effect of bending is shown with the areas of compression and stretching or tension. A line is drawn square across the pipe; the broken line shows its new position after bending. Fig. 178 shows the effect when the bend is pulled on a spring. The line just mentioned has moved through 90°; the effect of thinning and thickening is not surprising.

The solution to the problem is self-evident; the inside surplus must be worked round to the back or outside of the bend.

The bend to be made should be chalked on the floor or on some other surface so that a ready check is available. The job should be carefully thought out; if the bend is near one end of a longish pipe, the work will have to be done from one end. For the best results the farthest part of the bend will have to be made first.

Having marked the length of the bend on the pipe, a small indentation should be made at the first point of bending. Now the pipe should be gently warmed until a few drops of water are seen to dance vigorously on the surface. The hand, protected by a piece of felt or sacking, is now placed on the heated indentation, and the end of the pipe pulled up about 30°. The pipe will now be seen to have a section as in Fig. 176, Step 1.

The pipe is now supported as in Fig. 179; either as shown, or by slinging from a rafter or girder, or held by a mate or apprentice. The pipe must be supported so that the bend is just clear of the bench, otherwise the bend will tend to open. The dummy (Fig. 172) is now introduced, and, using the bench as a fulcrum, the inside of the bend is lifted by blows from the solder end. These blows are applied by rapid downward strokes of the other end. By watching the bend the blows can be directed at will. The

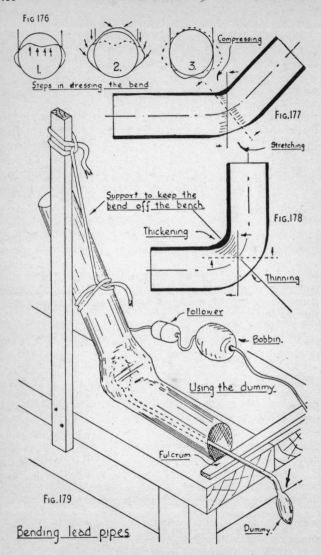

FIG 176

1. 2. 3.

Steps in dressing the bend

Compressing

FIG.177

Stretching

Support to keep the bend off the bench.

Thickening

FIG.178

Thinning

Follower

Bobbin.

Using the dummy.

Fulcrum

FIG.179

Dummy.

Bending lead pipes.

dummy is best made of steel, the solder end should be shaped as shown in the elevation (Fig. 172), but flattened in plan.

The section should now be as shown in Fig. 176, Step 2; that is, having the correct depth but extra width. This surplus should be dressed round to the back as indicated by the arrows until the section is as seen in Step 3. The operation is now repeated until the desired bend is achieved. These successive bendings will move in small stages towards the end of the pipe.

Using the dresser (Fig. 169) and the bending stick (Fig. 170), the bend should now be dressed as near as possible to its final shape. The ends of the pipe should be dressed on the mandrel (Fig. 175) to remove any indentations. Finally the bend should be finished to the true bore by means of a bobbin (Fig. 174). These are usually made from hardwoods such as lignum vitæ, boxwood, or beech. Before insertion the bobbin should be smeared with tallow to reduce friction. The bobbin is threaded on to a cotton rope, on which it slides and which has anchored upon it a metal weight or follower. Two forms are shown in Figs. 171 and 173. The first is usually of brass, the two halves being secured by a set-screw. A figure-of-eight knot is used to secure it. The second is of wrought-iron or brass. The rope is passed through it and looped out at the hole and knotted. Alternatively, the figure-of-eight is tied and the ends worked through from the side hole.

To use the bobbin, the end of the rope is passed through the pipe; this is followed by the bobbin which will ultimately stick at the beginning of the bend. This again is followed by the weight. Two men are needed for this operation. The craftsman, taking the leading end of the rope, should pull on it until the weight is close up to the bobbin; he should now take a firm hold on the rope about a foot from the pipe end. For greater purchase the rope can be hitched to the head of a hammer—this preserves the fingers. The assistant now taking the other end of the rope, draws the weight back about 230 mm (9 in). The craftsman now draws or snatches the weight sharply against the bobbin, which is driven along the bend. This operation is repeated until the bobbin has passed through. At the same time, the craftsman should keep an eye on the bend, and where necessary dress the bend on the bobbin.

Some plumbers make a practice of driving the bobbins by means of a stick. On the larger pipes this should be avoided,

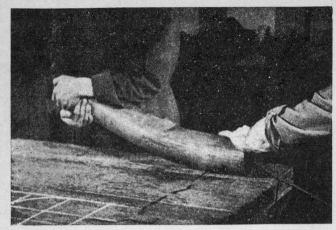

Raising the Throat of the Bend by means of a Dummy.

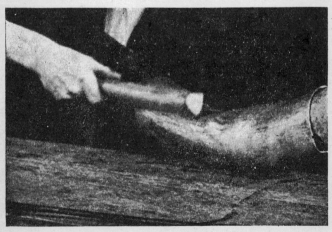

Dressing Surplus Lead to the Back of the Bend.

PLATE 6—LEAD-PIPE BENDING.

because driving makes progress on the outside of the bend, whereas pulling makes progress on the inside, particularly if heat is applied to the inside as the pulling proceeds.

If the bend to be made is an acute one, the inside of it should be well raised in dummying, and a maximum amount of lead should be driven to the back. The dummy should be used carefully and evenly and creases should be avoided, particularly in driving lead to the back from both sides. Where the bend is very near the end of the pipe, a small hand dummy (Fig. 168) may prove useful.

CHAPTER XV

COPPER-TUBE BENDING

ONE of the advantages in the use of copper tube in building is the ease with which it can be bent for almost any position. Skill in bending saves considerably in the cost of fittings for angles. Slow bends can be made, which will reduce resistance to flow, and a neater appearance can be achieved.

The effect of bending on a copper tube should be understood if the beginner is to succeed. When a tube is bent—as a simple trial will show—the back of the bend tends to become flattened, the inside or throat wrinkles, and the sides swell outwards. The pipe instead of remaining circular in section becomes elliptical, and although the distance round the pipe is the same, the cross-sectional area and the capacity of the pipe will be greatly reduced.

For this reason some way must be found to prevent this "collapsing" at the bend and to maintain the pipe in its original circular section.

The methods used are designed to support the walls of the tube whilst bending takes place. The support can be given either inside or outside. For inside support, some form of loading is used; for outside support we use a machine.

The materials most generally used for copper-pipe loading are:
(i) Steel springs; (ii) molten lead; (iii) pitch or/and resin; (iv) sand; and (v) proprietary materials.

Spring Bending

Springs can be used in bending pipes up to 50 mm (2 in) diameter. They are spiral springs which, being made from steel of rectangular section, present an outside flattened surface. Springs of round wire are also sometimes used. Copper tube or pipe is usually supplied in "half-hard" temper for reasonable rigidity. It can easily be annealed by heating to dull redness. Using springs, tubes up to 25 mm (1 in) diameter can be bent without annealing, but the larger pipes should be softened.

To anneal a copper pipe a large blow-lamp is needed, the normal lamp being too small, in view of the speed with which heat travels in copper. In heating a tube, the ends should be stopped with corks or paper, otherwise large quantities of air will be heated.

Springs are generally fitted with handles which can be removed where the spring must be used in the middle of a pipe. In this case a long rod fitted with a hook is used to extract the spring. In spring bending, the bend should be pulled a little "over-square" and then pulled back. By this means the slightly elliptical section is restored to the round and the spring can easily be extracted. To remove the spring, it should be turned so as to tighten the coils; this action reduces its diameter. It can then be pulled out.

With the larger pipes (50 mm—2 in—diameter), it will be necessary to anneal at first and again when the bend is half made. This is because the bending "work-hardens" the copper. In this event, the spring must be removed whilst the second annealing is done, and reinserted afterwards. A little grease on the spring may help. On no account should a dresser or mallet be used on the pipe whilst the spring is still inside, otherwise it may not be removable. To have witnessed this impasse is an awful warning. For the larger pipes, the bend former (Fig. 185) may be a great help.

Lead Loading for Bending

Lead loading is suitable for pipes up to 38 mm ($1\frac{1}{2}$ in) diameter only, because of the difficulty involved in melting out the lead afterwards. Before loading begins, the pipe must be thoroughly annealed for the whole length of the bend. If water is used to cool it off, the pipe must be properly dried before molten lead is poured in; otherwise steam will form and may blow the molten lead on to the operator.

The lead should be pure, soft pig lead or melted-down sheet scraps, care being taken to eliminate any solder. The lead should be very hot and should be poured into the pipe as quickly as possible to avoid the formation of air pockets.

Using this method, bends of very short radius can be made; up to 25 mm (1 in) as sharp as 2 diameters, and 32 mm ($1\frac{1}{4}$ in) and 38 mm ($1\frac{1}{2}$ in) pipes can be bent to $2\frac{1}{2}$ diameters radius. This way of measuring bends is illustrated in Fig. 183.

Small pipes can be bent across the knee; for larger diameters the best method is demonstrated in Fig. 180. The copper pipe should be held in the vice by means of cast-lead clamps. An iron or steel lever is required, and an iron sling, together with pieces

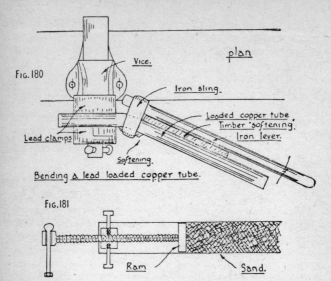

Fig. 180

plan

Vice.

Iron sling.

Loaded copper tube.
Timber softening.
Iron lever.

Lead clamps

Softening.

Bending a lead loaded copper tube.

Fig. 181

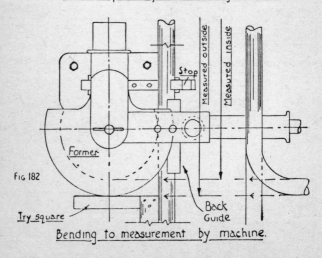

Ram

Sand.

Sand compressor for hot bending with sand.

Measured outside

Measured inside

Stop

Former

Fig 182

Try square

Back Guide

Bending to measurement by machine.

of softening. The pipe will bend at the point at which the sling is fitted, and which will therefore be moved in steps in forming the bend. A template or pattern should be made from a piece of stout wire or old lead pipe so that the bend can be checked in the vice.

To melt out the lead, considerable heat will be needed, and a powerful lamp should be used, or, if available, an oxy-acetylene flame.

Finally, molten lead is dangerous material to handle, and the pipe to be loaded should be fastened in a vice or held by means of large tongs. Where possible the plugged end should rest on the floor or have other support.

Sand Loading

In sand loading, the pipe is rammed solid with dry sand, and the sand must be perfectly dry, otherwise when heat is applied to the pipe steam may be produced and an explosion could occur.

One end of the pipe should be plugged by means of a tight-fitting slow-tapered wooden plug. The dry sand can be poured in through a paper funnel and well tamped down. The sand should reach within 40 or 50 mm ($1\frac{1}{2}$ or 2 in) of the top end, and a second plug should be driven in. As the bending tends to squeeze the sand out towards the ends of the tube with considerable force, the tube can be drilled and a screw inserted in each plug. The holes can afterwards be welded up or the ends cut off.

Where a fair amount of bending is done, a sand compressor of the kind shown in Fig. 181 should be made. Besides securing the ends, this compressor compacts the sand efficiently. The ram can be fitted with ends appropriate to each size of tube.

Small pipes can be bent cold, but most successful sand-loaded bending is carried out with a red-hot pipe.

Having loaded the pipe, the part at which bending is to take place should be heated to red heat. This can be done by using a powerful lamp, an oxy-acetylene flame, or in a forge. If a forge is used, the pipe must be carefully watched and revolved to get an even heating. In the process of heating, the sand may shrink a little, and the plug must either be driven in a little or the compressor screwed up.

The bend may be started by what is called "dumping." Here the pipe is held at an angle with the floor and heavily dumped by adding the weight of the body to the downward thrust. The bend should be inspected and reheated and bent still farther at the point decided upon. A template must be on hand, or a sketch should be drawn on the floor.

By sand loading bends of short radius can be made.

Where the bend is dented or crimped, a steel bobbin should be driven in, and the irregularities evened out by swaging with a special swaging tool or by means of a round-faced hammer.

Where a sand compressor is used, the screw should be a standard gas thread, so that it can be lengthened by the addition of pieces of gas pipe. By this means sand can be packed at a distant bend.

There is an excellent loading material on the market, which sets and melts at a temperature below that of boiling water. The advantage of this material will be obvious.

Machine Bending

The bending machine is designed to bend a pipe whilst preserving its true circular section. This is achieved by bending the pipe around a former wheel having a groove, whilst the back of the bend is supported by means of a semicircular back-guide (Fig. 182).

Bending machines are available on special stands or for fixing to a bench or plank. The former wheels and back guides are supplied for each size of pipe. Hand-lever machines are intended for pipes up to 50 mm (2 in) diameter, and geared and powered machines will bend the larger sizes.

Most bending-machine manufacturers recommend the annealing of pipes before bending, but half-hard copper pipe can be bent without much difficulty up to 32 mm (1¼ in) diameter.

Sometimes the inside or throat of a machine-made bend is found to be corrugated or crimped. This is due to the point of pressure, by the roller on the back-guide, being too far in advance of the bending point in the former wheel. This calls for a slight tightening of the screw on the handle which controls the position of the roller.

Setting Out Bends

Where parallel bends are needed, as in Fig. 184, only one of these can be made by machine because they differ in radius and all machine-made bends are the same radius. One of the bends must therefore be made by some form of loading. In most cases

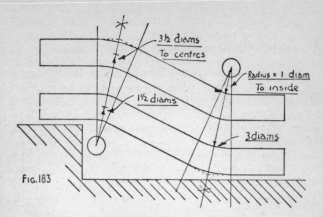

3½ diams
To centres

Radius = 1 diam
To inside

1½ diams

3 diams

Fig. 183

Setting out a pair of offset bends.

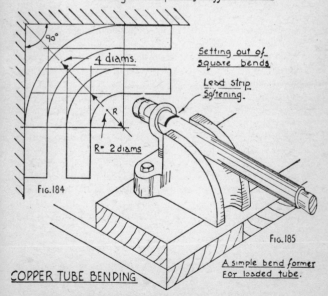

90°

4 diams.

Setting out of
square bends

Lead strip
Softening

R

R = 2 diams

Fig. 184

Fig. 185

COPPER TUBE BENDING

A simple bend former
For loaded tube.

the machine bend would be used outside because of its long radius. In the case in point, the inside bend is 2 diameters in radius, whilst the outer bend has a radius of 4 diameters, measured to a line in the centre of the bend.

To ascertain the radius of the bend to be made by hand bending, the machine bend should be placed against a square or in a corner, and the point of generation found (Fig. 184). The distance between the pipes can now be decided and the second bend marked out.

Fig. 183 illustrates a similar operation where parallel offsets are needed. Here the slow bends are found to have a radius of $3\frac{1}{2}$ diameters and the others a radius of $1\frac{1}{2}$ diameters (to the centre lines). The wide-radius bends would be made by machine and the others by methods of loading.

CHAPTER XVI

DOMESTIC HOT-WATER SUPPLY

The Cylinder System

IN providing a satisfactory supply of hot water, it is necessary to consider a means whereby water can be heated, stored, and replaced as it is used. The cylinder system has evolved as a means of fulfilling these requirements.

Its evolution can be traced and its principles understood by reference to the sketches in Figs. 186–188.

Fig. 186 shows how water will circulate by gravity. If a crystal of permanganate of potash is dropped into the tube at X and a flame is applied at A, the water will be seen to circulate in the direction of the arrows. In Fig. 187, a vessel is inserted at A, this acts as a boiler, and by providing a greater amount of heating surface, causes a much more rapid circulation of the water. To observe the effect more clearly, a little sawdust stained in waterproof ink is most effective. By including a large vessel at point B in Fig. 188, it will be seen that a large quantity of water can be gradually heated by means of the circulation. Fig. 188A shows how hot water can be drawn off at point C, and how it can be replaced by the provision of a supply vessel and pipe at D.

We have now the structure of the cylinder system of hot-water supply or of the tank system, in which the cylinder B is replaced by a rectangular tank.

The principles involved are the same, so that attention will be focused on the more usual cylinder system.

This is demonstrated in the line diagram in Fig. 189. In most houses the water is heated in a fire-back boiler (Figs. 189 and 215). It is usually made of 6 mm ($\frac{1}{4}$ in) copper-plate with brazed joints, but is sometimes of cast-iron or even of aluminium. In bigger houses a boot boiler (Fig. 214) is sometimes used in order to provide a greater heating surface and many ordinary boilers are provided with arched flues (Fig. 214) for the same purpose.

The Circulation Pipes

Hot water is conveyed from the top of the boiler to a point near the top of the cylinder by the "primary flow" pipe, the cooler water from the bottom of the cylinder being returned to

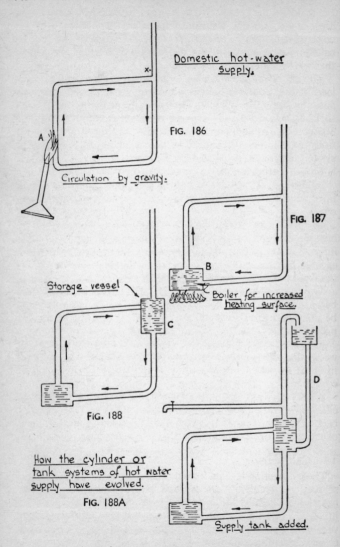

Domestic hot-water supply.

FIG. 186

Circulation by gravity.

FIG. 187

Boiler for increased heating surface.

Storage vessel

FIG. 188

How the cylinder or tank systems of hot water supply have evolved.

FIG. 188A

Supply tank added.

the boiler by the "primary return" pipe, which terminates within an inch or so of the bottom of the boiler. The circulators are connected to the boiler by means of boiler unions, sometimes called plumber's unions (Fig. 191), a dip-pipe being attached to that which takes the return pipe.

The unions are "made good" to the boiler by means of white lead or a good non-poisonous and reliable jointing compound and a gasket of hemp tow. They must be tightened down well so as to avoid leakage, using 220 mm (9 in) footprints or a 300 mm (12 in) Stillson wrench. It is well to remember that the fireplace may be built up before the water is turned on to the system. Particular care should be taken to see that the dip-pipe is on the correct union.

In deciding on which side the unions will be placed, attention must be given to the running of the pipes so as to avoid crossing over.

The range-fixer should be asked to provide a flue around the bottom and back of the boiler. This flue should be 50 mm (2 in) in depth, and not less than 150 mm (6 in) in width. If the depth exceeds 50 mm, cool air is often drawn through; the draught is thereby diminished, and inefficient heating results.

The pipes must be arranged so that they do not enter the flue, because of the harmful effect of flue gases, from which they must be protected. Much ingenuity is needed at times to effect this precaution.

Lead or copper may be used for these pipes, and it is well worth while to provide strong pipes in a situation where renewal involves the taking out of the fire range. Local by-laws must be consulted.

Black smoke stains are often seen where the circulators emerge from the chimney-breast. The pipes should be well bedded and the hole thoroughly filled throughout the thickness of the brickwork if this annoyance is to be avoided.

The Cylinder

The cylinder should be fixed as near as possible to the boiler, so that the circulators are as short as practicable, to minimise heat losses. Ideally, the cylinder should be placed alongside or just above the boiler, but in practice it is generally most convenient to fix the cylinder in the bathroom, where it can be used to heat an airing cupboard.

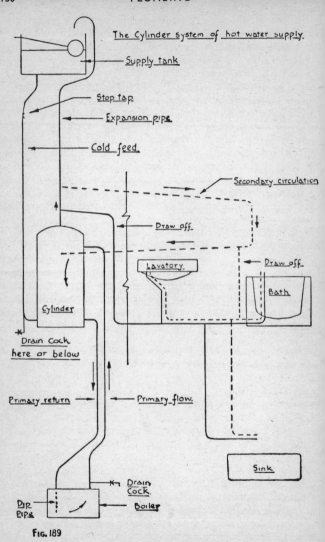

The Cylinder system of hot water supply.

Supply tank

Stop tap

Expansion pipe

Cold feed.

Secondary circulation

Draw off.

Draw off.

Lavatory.

Bath

Cylinder

Drain Cock
here or below

Primary return

Primary flow.

Sink

Drain
Cock.

Dip
Pipe

Boiler

Fig. 189

This storage vessel, like the boiler, is usually of copper, and is designed to hold about 136 litres (30 gallons) in a normal domestic system. The seams are brazed for greater security, and in choosing a cylinder reference should be made to British Standard Specification No. 699. For strength, the top of the cylinder is made convex and the bottom concave. This also obviates rocking on the base and gives an inverted funnel effect at the top, preventing the accumulation of air bubbles.

The connections to the cylinder are made in the positions shown in Fig. 189. As many hot-water systems have proved inefficient through faulty placing of these connections, a few important points will be considered.

The primary circulation pipes should be connected as shown, the primary flow being placed about 230 mm (9 in) from the top and the return leaving as near to the bottom as possible. Not uncommonly they are found placed close together near the bottom of the vessel. The weakness of this arrangement will be evident when it is seen that any hot water delivered at the cylinder has to make its way to the top of the vessel through a large quantity of cold water. Further, the successful circulation of water in the system depends on the difference in the temperature of the water in the two pipes. Where the pipes enter the cylinder close together, there is a great amount of mixing of hot and cold water and the circulation is generally sluggish.

The cold-water supply pipe should enter the cylinder near the bottom, and preferably in a horizontal direction. As the water in the cylinder should lie in layers of hot at the top, warm, tepid, and cold at the bottom—merging of course into each other—a cold supply pipe entering vertically will provide an upward jet, when hot water is being used, and the hot water available at the top will be cooled by mixing.

At the top of the cylinder there can be one or more connections. Here provision must be made for an expansion pipe which will discharge by means of an open end over the supply tank, and for a draw-off pipe to carry hot water to the taps. Most cylinders are, however, provided with only one union on top and the pipes are arranged as shown in Fig. 189.

There is need for a word of precaution here. It is often most convenient to carry the draw-off pipe from the taps to a position close to the cylinder and then vertically up a pipe board to an open end over the cistern. Then a short piece of pipe from the

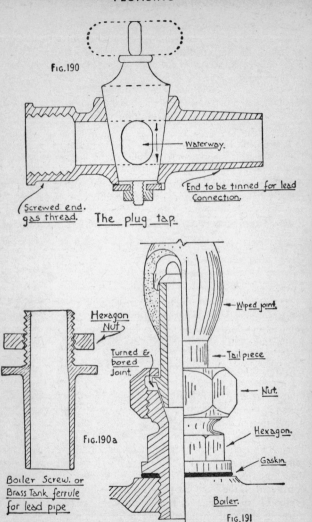

Fig. 190

Waterway.

End to be tinned for lead
Connection.

Screwed end.
gas thread.

The plug tap.

Hexagon
Nut.

Turned &
bored
Joint.

Wiped joint.

Tail piece.

Nut.

Hexagon.

Gaskin.

Fig. 190a

Boiler Screw. or
Brass Tank ferrule
for lead pipe

Boiler.

Fig. 191

The boiler union.

cylinder top union is branched into it. Care must be used to see that this short pipe rises from the cylinder to the pipe. Otherwise air which heating constantly expels from the water, and which accumulates at the top of the cylinder, will form an air-lock and so prevent the supply of hot water.

Valves

In general, the use of valves should be avoided where possible on hot-water systems.

It is useful, however, to place a valve or tap in the cold-water feed pipe immediately under the cistern. It should be a full-way type gate valve providing no impediment to the flow of the water, or of the plug-type as shown in Fig. 190. When this valve is closed, the hot-water supply is immediately stopped, all distribution pipes can be emptied and most repairs can be executed.

If it is deemed necessary to control the heating of the hot water in the cylinder, a valve can be inserted on the return circulator. In no circumstance should two valves be used, otherwise the boiler and pipes will be enclosed and an explosion might follow.

The Supply Tank

It is essential that there should be available a sufficient quantity of water to replace any drawn off the system. As a rough working rule, there should be enough to replace the water drawn for a bath, say about 90 litres (20 gallons). This tank or cistern is most commonly made from galvanised sheet iron of substantial thickness and according to local water regulations. The joinings of the sheet metal are welded or riveted.

In the average house this tank is fixed immediately above the cylinder in the top part of the bathroom airing cupboard, but sometimes, and particularly where greater head is required, it is installed in the roof space.

The connections to the tank are three, excluding the expansion pipe, which must be secured in such a way that its open end will discharge over the tank. The cold feed to the storage cylinder is connected in the handiest position in the bottom of the tank or preferably at a point as low as possible on the side. For the purpose a special brass ferrule (Fig. 190a) is used. It is provided with a shoulder, thread and a back-nut, and can be supplied prepared to take the joint to lead or copper pipe.

At a high level in the cistern or tank—both terms being

commonly used—connection is made to the cold-water rising main and to an overflow pipe. These details are the same as in the case of a w.c. flushing cistern and can be seen in Figs. 132 and 133. It should be arranged that the overflow shall be placed slightly lower than the ball tap, so that in case of overflowing the valve and the hole into which it is fitted are not submerged.

Fig. 201 illustrates a tank cutter, a tool specially designed for cutting holes in thin sheet metal. It is used in an ordinary hand brace, with care that the tool is held so that an even cut is made. Alternative types of cutter are available, one useful kind uses a circular piece of hack-saw blade. In emergency, the holes can be cut out by means of a small cold chisel and hammer, whilst the metal is held firmly on a lump of lead or hardwood, finishing off with a half-round file.

Where a brass fitting, such as an overflow ferrule (boiler screw) or a ball tap, has to be made good to a sheet-metal cistern, the hole should be no bigger than is necessary. The joint is made by means of a lead washer on the inside—against the shoulder— and a small quantity of a suitable jointing material. The lead washer provides a softening which will take up any inequalities in the surfaces whilst the jointing paste—or putty and paint— will make the connection watertight.

One final precaution: in the average domestic hot-water system there should be a margin of about 100 mm between the normal water-level and the overflow. This will allow the water to expand on heating without the nuisance of an overflow.

Secondary Circulations

Where fittings, such as sinks, baths, etc., are situated at a distance from the storage cylinder and a long draw-off pipe is needed, a large amount of cold water will need to be drawn before hot water is available.

This difficulty should always be avoided where possible.

To achieve this desirable end, a secondary circulation is used; this arrangement is pictorially described in Fig. 189. There it can be seen that the circulation is taken from a point in the expansion pipe and falls on its travel until it returns to a point on the side of the cylinder. Alternatively, where necessary, the secondary circulation may leave the expansion pipe—or the top of the cylinder—and rise to some convenient point on its circuit, then fall back to the cylinder as before. Where this is done, a vent or expansion pipe must be taken from the highest point to a position above water-level. Otherwise air will gather here and the circulation will be stopped.

The secondary circulation should be so contrived that it passes close to as many draw-off taps as practicable, the limitations to this aim are caused by obstructions to pipe runs such as doors and windows, and it can easily be seen that careful thought must go into planning such a scheme.

As hot water will be circulating close to all taps, no wasteful running off of cold water will be needed. It should be remembered that when a long draw-off pipe is used, it will be left full of hot water each time the hot-water tap has been used. This now cools off and the loss of heat in a day is considerable and very wasteful.

In order to conserve heat in the system, which is only produced by burning fuel, the secondary circulation should be insulated. The pipes can be boxed in and packed around with sawdust, slag-wool, spun glass or any other such non-conducting material. Or they may be wrapped with specially prepared strips of such material or covered in boiler-covering composition.

Trapped Circulations

Sometimes circumstances of layout are such that a trapped circulation cannot be avoided. A typical situation in which this problem will arise, occurs when the return circulator has to pass beneath a door and rise again to a point on the side of the cylinder, so forming a "trap." In order to overcome this resistance to circulation, the secondary flow-pipe must leave the expansion pipe at—or if taken off the top of the cylinder independently, should rise to—a height above the top of the cylinder not less than three times the depth of the trap. This should ensure an efficient circulation, but if it should be found sluggish, the height should be increased. .

Pipe Sizes

To determine pipe sizes in larger hot-water installations, calculations based on the quantity of hot water needed should be made, although previous experience generally provides satisfactory "rule of thumb" guidance. Such calculations are beyond the purview of this book, but they should be mastered.

Any system providing hot water is known as "domestic";

we must limit our concern to the average small-house installation.

In general, the back-boiler used measures about 250 mm (10 in) in length, is about 200 mm (8 in) from back to front, and 150 or 180 mm (6 or 7 in) deep. With an ordinary fire, a boiler of this size provides enough heating surface to keep the cylinder supplied with hot water.

The primary circulation pipes are normally 19 mm ($\frac{3}{4}$ in) diameter lead or copper pipes. They should not be less and, if the cylinder is a large one, 25 mm (1 in) circulation pipes should be used.

The expansion pipe must be equal in size to the circulation pipes, and is usually 19 mm ($\frac{3}{4}$ in) in diameter.

The cold feed-pipe is often specified as one and a half times the diameter of the largest draw-off. In practice, equally sized feed and draw-off pipes are found to be efficient. In an ordinary house system they are usually both 19 mm ($\frac{3}{4}$ in) in diameter.

There should be no restrictions in the supply pipe to a bath, so that a 19 mm ($\frac{3}{4}$ in) pipe to a 19 mm ($\frac{3}{4}$ in) tap is needed.

The branch draw-off pipes to sinks and lavatory basins are generally 12 mm ($\frac{1}{2}$ in) diameter to 12 mm ($\frac{1}{2}$ in) taps.

The cold-water rising main of 12 mm ($\frac{1}{2}$ in) diameter, being under pressure, is quite capable of replacing the water taken from the supply tank by gravity flow under such low "pressure head".

The overflow pipe and outlet should be sufficiently large to take the full output of the ball tap in case of failure. A 19 mm ($\frac{3}{4}$ in) pipe is mostly used, but a 25 mm (1 in) diameter overflow would give a more satisfactory margin of safety.

Cylinder Draining Tap or "Wash-out" Tap

To facilitate repairs to the cylinder, its connections, and any pipes below cylinder-top level, it is useful to provide a draining tap and run-off pipe. Some water authorities require this provision, whilst others do not and some will not allow it. In any locality the regulations should be consulted in order that the tap can be fitted in the required position.

In some areas the connection is made as in Fig. 189, where a tee-fitting replaces the elbow on the cylinder feed-pipe.

Elsewhere the drain-off tap is fitted near the bottom of one of the primary circulation pipes; the return pipe is generally sug-

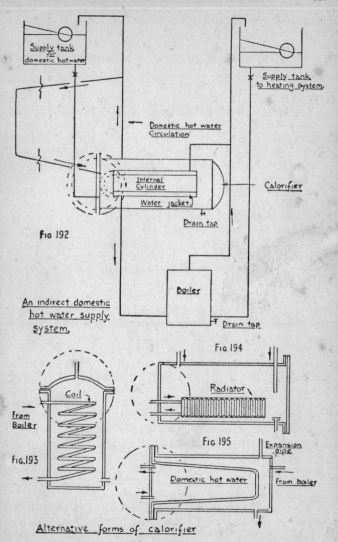

Fig 192

Supply tank for domestic hot water

Supply tank to heating system.

Domestic hot water Circulation

Internal Cylinder

Water jacket

Calorifier

Drain tap

Boiler

Drain tap

An indirect domestic hot water supply system.

Fig 194

Coil

From Boiler

Fig.193

Radiator

Fig 195

Expansion pipe

Domestic hot water

From boiler

Alternative forms of calorifier

gested, but as water finds its own level, there is little point in this. This arrangement is shown in Fig. 189 as the alternative.

The tap used for this purpose may be of screw-down or plug type, but gate-valves are also used. Once more the local regulations should be consulted.

The use of this provision is so rarely needed that many authorities regard it as unnecessary, and in this case it is a simple matter to empty the cylinder by means of siphonage. For this purpose a piece of small-diameter lead gas pipe or rubber tube is used, the water being run off into the bath or, if necessary, into buckets from the top union on the cylinder which has been disconnected. It should be remembered that the system can be drained down to this level by closing the stop tap on the feed from tank to cylinder and opening a hot-water tap.

Indirect Hot-water Systems

Where a hot-water heating system exists in a building, it is often expedient and economical to utilise the heating surface of the existing boiler.

It is very undesirable to use the hot water in the heating system for domestic hot-water supply purposes, because of the yellow discoloration from the iron pipes and boiler, and because of the harmful effects on this apparatus through constantly changing the water in them. Fresh water contains free oxygen which assists rusting and is driven off by heating. The constant addition of fresh water accelerates rusting and must be avoided.

Because of this difficulty, the heating surface in the heating boiler is used indirectly by means of a calorifier. This is a heating vessel—as the name implies—and consists of a cylinder in which there is some form of heater, heated by means of circulation pipes from the boiler.

Fig. 192 shows such an arrangement where an internal hollow cylinder is used within the domestic hot-water cylinder. Here the hot water from the boiler gives off its heat from the annular space of this internal cylinder and returns cool to the boiler. The domestic cylinder functions in the normal way. A secondary circulation of domestic hot water is shown.

There are many types of calorifier available. Alternative forms, suitable for medium-sized installations, are depicted in Figs. 193–195, where a coil, a radiator, and another form of internal cylinder is used.

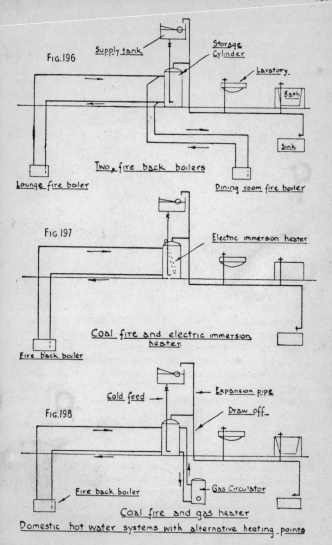

Fig.196

Supply tank

Storage Cylinder

Lavatory

Bath

Sink

Two fire back boilers

Lounge fire boiler

Dining room fire boiler

Fig.197

Electric immersion heater

Coal fire and electric immersion heater.

Fire back boiler

Fig.198

Cold feed

Expansion pipe

Draw off

Fire back boiler

Gas Circulator

Coal fire and gas heater

Domestic hot water systems with alternative heating points

It is very important that a ready means of access should be
provided to all calorifiers, so that the heating element can be
inspected and tested for leakage. The end is generally removable.

The Use of Independent Boilers

An independent boiler is a boiler which is not dependent on a
fire used for other purposes.

In many buildings large quantities of hot water may be needed
—and quickly. Even in some houses there are no open fires to
which a boiler may be fitted.

In such situations an independent boiler is often used in conjunction with a calorifier. However, the calorifier can be dispensed with if a Bower-Barffed boiler is used. The Bower-Barffing
treatment is applied by treating the red-hot cast iron with super-
heated steam, and so giving the metal a protective coating of
carbon which prevents corrosion in use. Where a boiler so treated
is used, circulators can be arranged directly between boiler and
cylinder. There is no dip-pipe, of course, in an independent boiler,
a return inlet being provided on the side near the base. Fig. 213
shows a section through a typical boiler.

Heating from Alternative Points

Figs. 196–198 show diagrammatically how domestic hot water
may be heated by alternative means. Most people having a hot-
water system utilising a fire-back boiler, have experienced the
difficulty of heating water for a bath in summer weather when
fires are not being used. Some are compelled to live all winter in
one room because the boiler is there.

Fig. 196 shows how two back-boilers may be used, so that two
rooms are always usable. Here one pair of circulators can be
simply branched into each other. Care should be taken that
"swept" junctions are formed so as to provide a minimum of
friction. Right-angled junctions must be avoided.

Fig. 197 illustrates the use of an electrical immersion-heater as
an alternative. This is a ready way of heating water for short spells
or in summer. The installation of the heater is described below.

Fig. 198 introduces a gas circulator as a means of boosting
or as an alternative to the coal-fire. Many types of circulator are
available. They can be thermostatically controlled and are
economical in use.

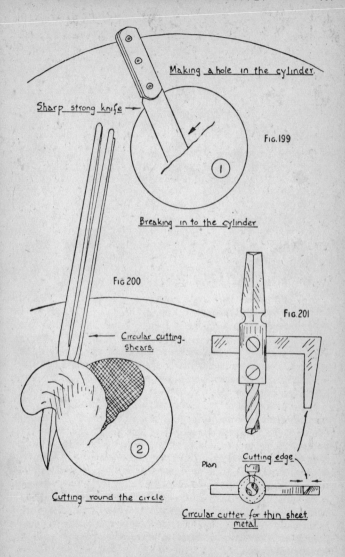

Making a hole in the cylinder.

Sharp strong knife →

Fig. 199

Breaking in to the cylinder

Fig. 200

← Circular cutting shears.

Fig. 201

Cutting round the circle

Plan

Cutting edge →

Circular cutter for thin sheet metal.

Installing an Electrical Immersion-heater

The fixing of an immersion-heater is the work of a plumber, though a competent electrician should make the electrical connections.

Most commonly, immersion-heaters are of the 2 or 3 kW type, with or without thermostat, and are fixed in the top of the storage cylinder diagonally downwards. Some are fitted with a surrounding, open-ended cylinder or sleeve, which compels circulation within the storage vessel. This is to be recommended.

There is much to be said for the use of two smaller heaters, inserted in the side of the cylinder, one near the top and the other at the bottom. They should be wired so that they could be used separately or together. The advantage in this arrangement is that, when a small quantity of hot water is needed, say for washing-up, the upper heater can be used. When a large quantity is needed quickly, say for a bath, both can be used, and lastly, where warm water is generally needed or hot water later in the day, the lower heater can be switched on.

Whichever method is preferred, the fixing to the cylinder is the same.

Inspection of an immersion-heater will show that the element is made to screw through a heavy internally threaded brass ferrule which is provided with a flange. It can be seen in section in Fig. 203.

This ferrule must be inserted and secured in the top of the cylinder, so that the heater can be screwed into it, and tightened down on to a fibre washer. For the purpose of tightening, a hexagonal nut (Fig. 202) is provided.

The outside diameter of the body of the ferrule should be carefully measured to ascertain the size of hole needed in the top of the cylinder. Next a method of cutting the hole must be devised. Often the method depends on accessibility, and indeed sometimes the cylinder will have to be disconnected and removed. In general, however, the fitting is done *in situ*. If the surface at the point decided upon is reasonably flat, a tank cutter may be used—given sufficient headroom for a brace. Alternatively the hole may be cut in the manner shown in Figs. 199 and 200.

An incision is first made by means of a sharp hacking knife, cutting as soon as practicable in a direction parallel with the surface as shown. Next a pair of shears designed for circular

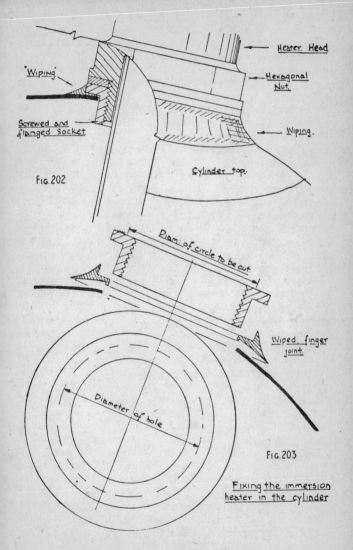

Heater Head

Hexagonal Nut.

Wiping.

Wiping

Screwed and flanged Socket.

Fig. 202

Cylinder top.

Diam. of circle to be cut.

Wiped finger joint.

Diameter of hole

Fig. 203

Fixing the immersion heater in the cylinder

cutting should be used (Fig. 200). To facilitate cutting, it is helpful to remove the pipe from the cylinder top union. In many cases,
where a drain cock is not fitted to the cylinder, this will already
have been done in order that the water-level in the cylinder could
be lowered by siphonage. Everything considered, this method of
cutting is generally the easiest.

The surface around the hole should be carefully straightened
and freed from any waves or kinks.

The underside and edge of the flange should now be tinned carefully, so that no part has been missed, and care taken that no
solder has been allowed to adhere to the threads. Tinning is dealt
with in Chapter XI on Soldering. The top of the cylinder around
the hole must also be tinned to a width of about 25 mm (Fig. 203).

The top surface can be—and must be—thoroughly cleaned by
means of coarse emery cloth, and in order that the solder shall
not spread widely over the surface, its limits should be defined by
painting the part outside the circle with plumber's black, to which
the solder will not stick. When the black has dried, the cleaned
copper should be smeared with a resinous or any other non-
corrosive flux. The solder tinning can be applied by means of
either a heavy copper bit or the blow-lamp flame. Perhaps the
best way is to use a hot copper bit, kept heated by playing the
blow-lamp flame on it. Tinning by flame needs good judgment
with most fluxes and, except by the highly skilled, should be
avoided.

The tinned ferrule should now be fitted in position and the
finger-wiping made, using plumber's solder of 2 parts lead to
1 part tin. The heater can now be screwed in and tightened, and
the system can be filled up again. The completed job is shown
in Fig. 202.

CHAPTER XVII

HEATING BY HOT WATER

THE term central heating applied to the heating of domestic and other buildings indicates that the whole of a building is heated from a central source. Usually an independent boiler, fired by solid fuel, gas, electricity or fuel oil.

The boiler is generally placed at the lowest available point in the building, having regard at the same time to convenience of stoking and delivery of fuel.

Boilers

The boiler may be one of a number of types. It may be a solid one-piece casting, rectangular in form; it may be sectional; or it may be conical in shape and of wrought or cast iron. For smaller systems, the first and last-named types are both cheap and suitable. The sectional boiler has the advantage of the possibility of added sections should more heat be needed subsequent to initial installation.

Sectional and shell type boilers are almost invariably used for bigger installations. The former are cast iron and can be built up *in situ*, whilst the latter are usually of the "packaged" type, having all auxiliary components together with the boiler assembled as one unit ready for erection. Maker's catalogues should be consulted for details of different boiler types. A section through a solid fuel small domestic boiler is shown in Fig. 213.

Design Considerations

In general, heating system should be designed so that the water will circulate by gravity (as described in Chapter XVI, Fig. 186). In some installations, circumstances are such that a pump or accelerator must be used to achieve a satisfactory circulation. This should be avoided if possible.

When designing a heating system for a large building, it is usual—in the interest of economy and to ensure efficient heating —to first calculate how much heat will be needed to maintain the building at the desired temperature. Then the size of the boiler

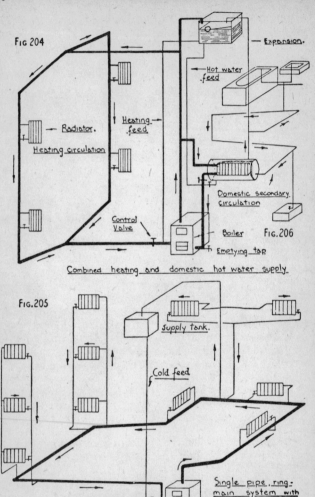

Fig 204

Expansion.

Hot water feed

Radiator.

Heating feed

Heating circulation

Domestic secondary circulation

Control Valve

Boiler   **Fig. 206**

Emptying tap

Combined heating and domestic hot water supply

Fig. 205

Supply tank.

Cold feed

Single pipe, ring-main system with risers

Control valve

Central heating systems.

and the amount of pipe and radiator heating surface required to give out this heat will be estimated. For small systems, "rule-of-thumb" methods and past experience are generally a sufficient guide.

The most commonly used layouts are illustrated by line diagram in Figs. 204–208.

The overhead drop-feed system (Fig. 208) shows how the hot water from the boiler is carried as high as possible in the building, from where it falls in cooling, through the various branch pipes and radiators, back to the boiler. In this type of system, the maximum amount of "circulating head" or pressure, would be obtained.

Circulating Head or Pressure

In any gravity system of heating (i.e. no pump), circulating head is extremely important. Briefly it is due to the difference in weight of a given volume of water in the flow and return circulators. This factor governs the speed of the circulation, and it should be borne in mind that the rapidity of the circulation will determine the amount of heat which will reach a given radiator.

It will readily be seen that if a pipe were taken from a boiler and carried around a room horizontally and back to the boiler, no circulation would take place—provided the boiler connections were on the same level. If one pipe were connected to the top and the other to the bottom of the boiler, a slow circulation would be found to exist. If instead a pipe leaving the top of the boiler were to be taken to the top of the room, allowed to circulate around the room and return to the bottom of the boiler, a rapid circulation would be evident.

This principle should be borne in mind in the design of any heating system.

Fig. 204 shows how this is utilised in a small installation, where incidentally, the domestic hot water is heated by means of a calorifier.

Two other systems in common use are shown in Figs. 207 and 208. In Fig. 207, the two-pipe system shows a method of heating where it is most convenient to have the main circulation in the basement, with flow and return "risers" branched into the appropriate flow and return main circulators. There is some advantage in this layout, in that a good circulating head is maintained

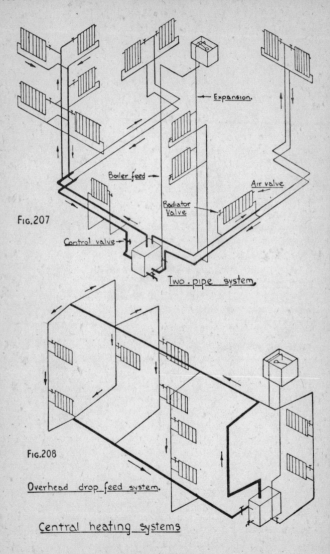

Fig.207

Expansion.

Boiler feed

Radiator Valve

Air valve

Control valve

Two-pipe system

Fig.208

Overhead drop feed system.

Central heating systems

by returning all water which has given up its heat in radiators to the main and separate return pipe. The other systems shown have the disadvantage that cooled water is returned to the main circulation which still has to supply hot water to other radiators.

Most heating installations are a compromise between what is ideal and what is practicable, in consideration of possible pipe runs, radiator and boiler positions, and the shape of any particular building.

There are a few rules which should be observed.

(1) The boiler must be big enough for the job it has to do.

(2) The pipes should be of a size sufficient to convey the volume of water required to give up the needed heat.

(3) Pipe runs should be as direct as possible.

(4) A system should be planned to provide enough heat in the worst possible conditions.

(5) The occupants of rooms should be given a means of controlling the heat output in the room by means of valves.

From the point of view of heat control, one valve only is needed and this can be fixed on either pipe; although in practice it is usually fitted to the return or lower pipe where it is out of the way. It is useful to fit a valve on each pipe so that radiators can be taken out for repair without having to run off the whole of the water in the system.

Erecting and Fixing Boilers

No difficulties exist in the fixing of independent boilers. A good level concrete foundation is needed, and common-sense fire precautions should be observed.

Sectional boilers can be almost any size to as much as 5 or 6 tonnes weight. They are usually delivered in separate parts requiring assembly. Heating by hot water is traditionally plumber's work, but in latter years specialist heating engineers have evolved, and it is true to say that larger heating jobs have become mainly their concern. However, a great deal of domestic heating and other smaller systems are carried out by the plumber; some of this work involves the use of sectional boilers.

The stand should be fixed on the level concrete foundation, and it is a good plan to raise the stand either on a course of brick-work or on a concrete curb of similar height. This gives a deeper ash-pit, which allows a considerable accumulation of ashes before the draught is impeded.

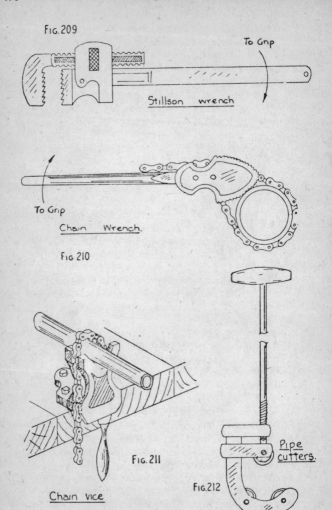

Fig. 209

To Grip

Stillson wrench

To Grip

Chain Wrench.

Fig 210

Fig. 211

Chain vice

Pipe cutters.

Fig. 212

It is advisable to assemble the sections on the floor rather than on the stand, so that it will not get pulled about and maybe damaged. If, however, the boiler is a heavy one, it can be assembled on the stand if care is taken. The front of the stand can be removed for assembly of the front section.

The sections may not meet closely and should not be forced together by undue effort. A special spanner is provided, and is long enough to give the necessary purchase in tightening. Under no consideration should a piece of piping be used to give extra leverage.

The sections should be assembled from the back in order of the numbers painted on them. First nipple holes should be carefully cleaned out by means of a piece of cloth or waste, the nipples should be cleanly wiped, and painted over with a little thin red lead and boiled linseed oil or a suitable jointing paste. The nipples should then be carefully centred, and the sections moved into position, using a pinch bar in the case of heavy sections. The bolts should be tightened to pull the sections together by working diagonally; that is, top right and bottom left, and so on.

A quantity of boiler cement is provided by the manufacturers for making the boiler smoke-tight. It should be used to seal the spaces between the sections, to make the sections good to the stand, and to joint the flue pipes.

The flue pipe for any type of boiler must be carried above surrounding buildings, and in any case to a height sufficient to provide a good draught. The aim should be a powerful draught which can be readily controlled by means of a damper or air inlet beneath the fire-bars.

Boiler Covering

All boilers should be insulated for efficiency, using a low conductivity material. Some boilers have a special metal sheet cover supplied, which is fixed in position, and the space between filled with a low conductivity material.

Where the boiler is to be covered with boiler composition, it should be mixed with cold water to a thick paste. Then it should be thrown on in small lumps until the surface is covered. When this coat is dry a further thin coating should be added and trowelled off smoothly. A total covering of about 38 to 50 mm ($1\frac{1}{2}$ to 2 in) is recommended, and will add nearly 10% to the

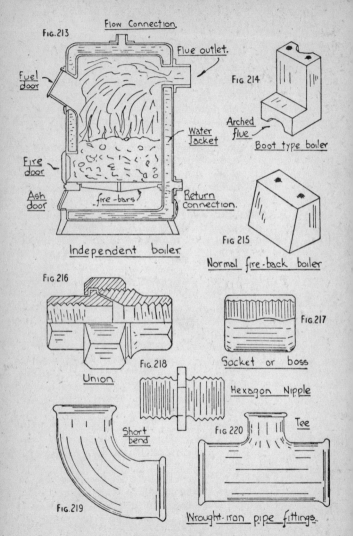

Fig. 213

Flow Connection.

Flue outlet.

Fuel door

Water Jacket

Fire door

Ash door

fire-bars

Return Connection.

Independent boiler

Fig. 214

Arched flue

Boot type boiler

Fig 215

Normal fire-back boiler

Fig 216

Union.

Fig. 217

Socket or boss

Fig. 218

Hexagon Nipple

Short bend

Fig 220

Tee

Fig. 219

Wrought-iron pipe fittings.

efficiency of the boiler. It should be remembered that any heat escaping outwards is not heating water.

Safety Valves

Whenever possible a boiler should be provided with an open-air vent pipe in preference to a safety valve; where this is not easily possible, a safety valve is provided in the top of the boiler. A suitable type of safety valve is illustrated in Fig. 222. This is of the dead-weight type, where weight rings are added to just more than counteract the pressure in the system due to head of water, i.e., the vertical height of the water level in the supply tank above the seat of the valve. The only trouble arising from this type of valve is due to grit getting on to the seating. For this reason the valve is best left undisturbed, or if it is moved it should be cleaned.

Fig. 221 shows an automatic draught control. It consists of a thermostatic valve operating a lever on an adjustable slide-scale and a chain, which raises or lowers a damper door. This valve is very sensitive, because the liquid enclosed in the capsule expands and contracts considerably with small temperature change. A pre-decided boiler temperature is fairly easily achieved and is steadily regulated.

A boiler thermometer (Fig. 223) is useful in checking on boiler performance. A tapping is provided on the top of the boiler for this purpose.

Fig. 213 is a section through an independent boiler suitable for a small heating plant or for domestic hot-water supply. It consists of a rectangular fire-box separated from the ash-pit by the fire-bars and surrounded by a water jacket. It presents a maximum amount of heating surface to the fire without the introduction of complicated water tubes.

The Cold Feed

The cold-feed pipe should not be taken directly to the boiler, but should enter the return circulator near to the boiler.

The Supply Tank

In fixing the ball tap in the tank, it should be remembered that the expansion due to a comparatively large amount of water

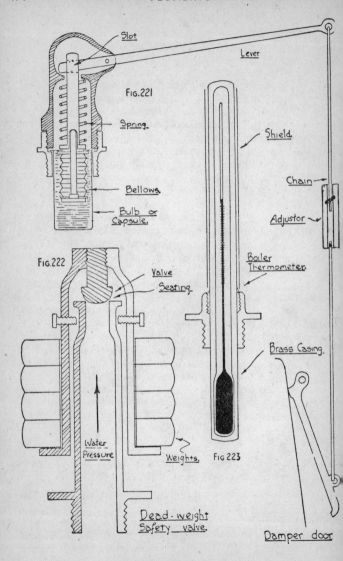

Slot

Lever

Fig. 221

Spring.

Shield

Chain

Adjustor

Bellows.

Bulb or Capsule.

Fig. 222

Valve Seating

Boiler Thermometer.

Brass Casing.

Water Pressure

Weights.

Fig 223

Dead-weight Safety valve.

Damper door

being heated will cause the water-level to rise. To accommodate this expansion, the ball tap is generally fixed well down in the tank, and a horizontal type of tank is to be preferred.

Pipe Fitting

Some of the tools used in pipe fitting are shown in Figs. 209–212. A portable vice (Fig. 211) is needed to hold the pipes, both for threading and assembly. The three-wheeled pipe cutters (Fig. 212) are a useful means of cutting pipes. In order that cutting should be quick and the tool unimpaired, the cutting must be constantly lubricated. Suitable wrenches are depicted; the Stillson wrench is a well-designed tool and handy in use (Fig. 209). The chain wrench (Fig. 210) is an older tool but still efficient.

For heating, a lap-welded, heavy, mild-steel pipe is used; this is painted red. Light and medium quality pipes are painted brown and blue respectively and are used for other services such as gas.

The thread used is the standard gas thread. The tapered dies used in the screwing stocks produce a slightly tapered thread, which tightens in the fittings. In recent years, stocks and dies have vastly improved in design, and are now easily operated and efficient tools. Before threading the end of a pipe, care should be taken that the burr produced by the cutters on the inside of the pipe is removed by means of a reamer.

To secure an effective joint the thread on the end of the pipe is smeared with jointing paste—preferably non-poisonous—and firmly tightened into the thread of the fitting.

Fittings are available for almost every purpose; a few are shown in Figs. 216–220. Bending is not much needed, in view of the diversity of fittings, but it still has its usefulness in some situations. Bending can be carried out easily by machine where the radius is suitable, and by forge-bending where necessary. The setting-out of machine bends is illustrated and described in Chapter XV on Copper Tube Bending.

The most important skill involved in pipe fitting is that of accurate measurement, and of estimating the take-up of fittings—so important, as failure on this score will invalidate accuracy.

Pipe Fastenings

Because pipes expand when heated, they must be allowed controlled movement. In general, then, pipes mostly require

Fig. 223A. Boiler with automatic mixing valve and pump.

support. Where long pipe runs occur, roller brackets are needed. Fortunately iron pipe has great rigidity and brackets or clips can be well spaced. In ordinary average-house rooms mere passing through walls, ceilings, or floors is often sufficient fastening.

Radiators

Most radiators are sectional; they can be extended and damaged sections can be replaced. A trade catalogue will readily indicate the many different patterns of radiator which can be had.

The majority of radiators have left- and right-handed malleable nipples connecting the sections together. This is a very useful arrangement, but it calls for some care on the part of the dissembler. To remove a section of a radiator, the plugs and bushings must first be removed from the ends. Now a special tool, designed to grip the ribs formed on opposite sides of the inside of the nipple, is inserted from the end. The depth to which it must go having already been chalked on it by trial against the outside of the radiator. The air-cock tapping is made on the return end of a radiator, which is left-hand threaded inside. A moment's thought will show that an anti-clockwise turn will screw the nipple out of the left-hand threaded end section and also out of the right-handed second section. Working from the other hand, the reverse will be the case; that is, a clockwise turn should be used.

When assembling radiator sections, the nipples should be cleaned and smeared with a good jointing paste. The thin composition washer should be fitted and the sections brought together, with care that they are in alignment. The best way is to lie the sections on two parallel planks.

When coupling up radiators in a heating system, it is most efficient to supply hot water at the top, and take the return water from the bottom at the opposite end. In this way the convection principle is utilised and more positive circulation is obtained.

An air-cock is fitted to each radiator so that pockets of air can be freed. These cocks are used when the system is being filled with water. The usual practice is to let out the air floor by floor, beginning at the lowest. Each air-cock should remain open until water appears, when it can be closed. Periodically the radiators should be freed of air which may have collected at the top.

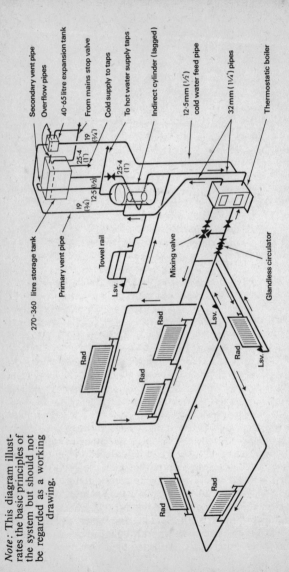

Secondary vent pipe
Overflow pipes
40·65 litre expansion tank
From mains stop valve
Cold supply to taps
To hot water supply taps
Indirect cylinder (lagged)
12·5mm (½") cold water feed pipe
32mm (1¼") pipes
Thermostatic boiler

270·360 litre storage tank
Primary vent pipe
Towel rail
Mixing valve
Glandless circulator

19 (¾")
25·4 (1")
12·5 (½")
25·4 (1")
19 (¾")

Lsv.
Rad
Rad
Rad
Lsv.
Rad
Rad
Lsv.
Rad

Note: This diagram illustrates the basic principles of the system but should not be regarded as a working drawing.

Fig. 223B. Typical small-pipe central heating system.

An air-cock is usually fitted with a loose key which fits on to the square end of the spindle.

Small-bore Heating Systems

After the Second World War, the demand for central heating in private houses grew. However, whilst gravity circulation was acceptable for larger buildings, the installation of this system in small houses proved very expensive and presented the installer with many problems—e.g. the difficulty of concealing large and unsightly pipes, damage to existing floors and walls, and, in new houses, the provision of certain changes in structural work, such as the notching of joints, the laying of pipes in special ducts, etc. The regulation of the heating system also proved difficult, especially on mild days in the winter if the domestic hot water temperature was to be maintained.

The introduction of the silent, small, electrically driven pump or circulator, which is glandless thus eliminating periodical attention and leaks, made modern small-bore systems possible (Fig. 223C shows a typical pump). This "silent" pump removed the objections to a pumped system for private house installations, previous pumps having been very noisy.

Fig. 223B shows diagrammatically the basic principles for a combined heating and domestic hot-water supply of a small system. The pipes are generally of 12·5 mm ($\frac{1}{2}$ in) bore in the individual loops and of 19 mm ($\frac{3}{4}$ in) bore in the common flow, and return near the boiler. They are either copper or black iron, and the hot water is forced through the circuits by the small electric circulator. An automatic mixing valve, either controlled by the outside temperature automatically or set by hand as required and fitted with a boiler by-pass connection, enables cooler water from the return main to be mixed with the hot water flowing from the boiler to the radiator system. This enables the heat emission from the radiators to be reduced in mild weather without a consequent reduction in the water temperature in the flow of the boiler. Thus it is possible to maintain the water in the boiler at a high temperature and hence a constant supply of hot water for the domestic hot-water system.

The indirect cylinder is connected to the boiler, usually by an independent gravity circuit to separate connection—pipes of 25 mm (1 in) bore—and thus is unaffected by the electrically driven

Fig. 223C. A typical pump.

pump (constant supply temperature for the domestic hot water can be obtained). By stopping the circulator, it is still possible to obtain hot water for domestic use without the central heating being run (see Fig. 223B); this is very important in the summer, and during warm periods in the spring and autumn.

Fig. 223A shows a typical small domestic boiler with the automatic mixing valve and pump installed externally to the boiler. With new installations the boiler, the mixing valve, the circulator and controls are in the common boiler casing. The boiler can be either coal, gas or oil fired.

Today, the radiators are usually the panel type made from pressed steel. They are either single or double panels. As regards heat output, sizes, etc., you should refer to the individual catalogues of the makers or suppliers.

CHAPTER XVIII

SANITATION

A MAJOR factor in public health in the past fifty or more years has been the contribution made by the plumber in the field of sanitation. That epidemics are almost unknown is in itself a testimony to this truth. The principles of good sanitary work are now well established and secured by public health law and local by-laws. Care should be used to see that installations conform with these statutory requirements. Our more immediate consideration is with the quality of workmanship.

Sanitary Fittings

Sanitary fittings are those appliances used in the collection and disposal of human and domestic waste products. They comprise commonly: water-closets, lavatory basins, baths, sinks, and urinals. The less common slop-sinks and bidets need not concern us.

Water-closets

This important appliance has evolved slowly from very crude beginnings to the well-designed, almost self-cleaning, present-day form.

We can see in sanitary museums the old hopper types, the ancient valve closet, the wash-out and other obsolete and insanitary appliances. We can examine the wash-down w.c., and see how its shape has gradually improved until it has become practically self-cleansing. The materials used have improved and the surfaces are more durable. We can see the greater regard for scientific principles, as in the application of siphonage in flushing.

The wash-down w.c. is depicted in section in Fig. 224, and attention is drawn to important points in design. There should be a large area of water surface in the bowl. In particular it should extend as far as possible from back to front. The back should give a vertical line as shown. The water seal must not be less than 50 mm. The flushing rim should be designed to distribute the flush over every part of the internal surface.

Wash-down w.c.s are available with P or S trap outlets, the

S trap being mostly used in ground-floor positions and the P
trap above ground-level.

The foregoing points should all be borne in mind when choosing
a w.c. pedestal.

Flushing cisterns of the siphonic principle must be fitted.
They must be designed to be "water-waste preventing," to conform with by-laws. (See Figs. 132 and 133.)

W.C. pedestals are made in glazed fireclay or in vitreous china,
the former are used mostly for outside conveniences. They can
be had in many colours to match interior decoration or baths
and lavatory basins.

Siphonic W.C.s

Siphonic-type w.c.s are quite an old idea, but in recent years
simplified forms have been put on the market, forms less clumsy
and involved. Fig. 225 is a line diagram demonstrating one such
appliance. The pedestal has an S trap and is fitted with a brass
shoe of 64 mm (2½ in) diameter; to this is attached not less than
1 m (3 ft) of 75 mm (3 in) diameter lead or copper soil pipe. The
operation is as follows: When the cistern is flushed, the length of
75 mm pipe is quickly filled with water and becomes the long
leg of a siphon. The short leg is that part of the trap which
descends to the pedestal. Once this siphon is filled with water,
siphonage takes place, and the contents of the basin are forcibly
removed. The remainder of the flush serves to refill the trap of
the pedestal.

The same principle is used in the siphonic w.c. seen in Fig. 226.
Here a change of pressures is brought about. The flush displaces
some of the water in the first trap, which in falling down the short
pipe A–B, causes a reduction of pressure at A and an increase at
B. It will now be observed that a state of unbalance exists in each
trap, with the excess of pressure acting in a direction from
pedestal to soil pipe. This force is enough to start siphonage.
When the pipe A–B becomes fully charged with water it forms a
siphon and the contents of the pedestal are removed almost as a
solid plug of water. The seal is remade by the remaining water at
the end of the flush.

This type has been much improved by having the second trap
built on to the back of the pedestal on the same level. This makes
it more convenient for fixing, as formerly the lower trap was
fixed in the room below.

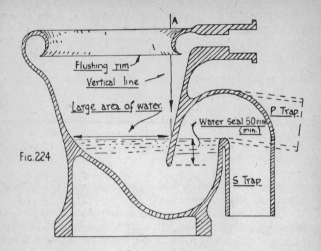

FIG. 224

Flushing rim
Vertical line
Large area of water.
Water Seal 50 mm
(min.)
P Trap
S Trap

Important points in the design of a wash-down
W.C. pedestal.

FIG. 225

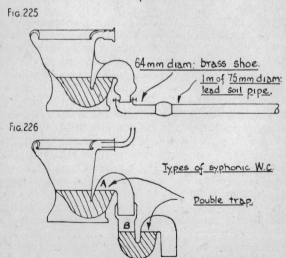

64 mm diam: brass shoe.
1m of 75 mm diam:
lead soil pipe.

FIG. 226

Types of syphonic W.C.

Double trap.

A

B

The siphonic w.c. is greatly to be recommended where a "low-down" cistern is used, and the flush has little "head" to give it force.

Fixing a Water-closet

Where an outside w.c. is to be installed, it will almost invariably stand on a solid floor; usually concrete. In this case the socket of a drain pipe or bend is brought up through the floor so that the end of the pedestal S trap rests on the shoulder inside. The distance of the drain inlet from the back wall will depend on the type of cistern to be used. Where an overhead flushing cistern is to be fitted, the pedestal can be quite near to the wall provided the flush-pipe bend can be accommodated. Such an arrangement can be seen in Fig. 247.

On a solid floor, the w.c. pedestal should be well pressed down on to a bed of soft cement; the cement being squeezed out on the inside to form a ridge which will resist any sideways thrust. The cement bed serves to take up any inequalities in the bottom of the pedestal and in the surface of the floor; so that the pedestal rests solidly. It is unnecessary to drill the floor and screw the pedestal down.

The joint to the drain should be made with cement with a little sand added to prevent shrinkage. The pedestal should be carefully centred in the drain socket, a ring of tow should be carefully packed in, and the socket half filled with the cement. When this has set, the socket can be filled and neatly trowelled off.

Figs. 247 and 248 illustrate the fixing of a wash-down w.c. on a wooden floor. Fig. 247 shows the flush pipe from an overhead flushing cistern. In Fig. 248 a "low-down" suite is diagrammatically presented. As above, this factor determines the position of the pedestal on the floor, and calls for careful measurement. It must be kept in mind that the flush pipe must fit up to the shoulder in the flushing arm of the pedestal.

On a wooden floor, the pedestal should be bedded down on putty. The floor should be marked around the pedestal by means of a pencil, and a coat of paint applied to the boards as well as to the underside of the pedestal before the putty is applied. The putty is merely a softening, and the pedestal is screwed down as firmly as possible by means of countersunk screws in the holes provided. A precaution should be observed here. A leather or similar washer should be fitted behind the head of the screw so that the porcelain will not be damaged, and brass screws should be used.

Generally the pedestal is connected to the end of a lead bend, which is passed through the wall. But sometimes the arm of a cast-iron junction is passed through.

Suitable joints to a lead bend are given in Figs. 231 and 232, using either a specially prepared brass thimble or a "Mansfield" socket. In the case of the brass thimble, it will be noticed that the lead pipe is made to pass through the thimble, and is turned over on the inside of the socket. Whenever brasswork is used in soil pipes, it must always be protected from the damaging acids to be found in "soil," i.e., human waste products.

These joints are made with cement and tow. Contrary to not uncommon practice, putty should not be used for the purpose, as leakage often occurs through pedestal movement, inevitable on a wooden floor without additional support. The pedestal needs the added holding effect of a rigid joint. It is often reasoned that putty should be used to facilitate repairs. This is not a very sound argument, as the need to remove the pedestal seldom occurs. The leakage is often too slight to be noticed, but is a constant cause of smells. The pedestal should on no account be jointed directly into the end of the lead soil pipe.

Alternative flush-pipe connections are shown in Figs. 227–230. There are many patent connectors available, all of which are useful. The "sunflower" connector (Figs. 229 and 230) is most useful in that it can be made on the job from an odd piece of scrap-lead. The method of cutting is given in Fig. 230. When the "sunflower" has been made, it can be secured to the flush-pipe by means of a finger wiping and the joint made with putty and paint. The soldering is not entirely necessary if the hole in the "sunflower" is a good fit on the pipe. The ancient rag-and-putty joint, though effective for watertightness, is an untidy and insanitary method, and should never be used. Modern low-cistern w.c. suites are usually supplied with patent joints for connecting the enamelled-iron flush pipe to the pedestal—utilising rubber rings instead of putty.

Lavatory Basins

There is some confusion in these days, when the water-closet has assumed the name of a washing vessel, and is so often called

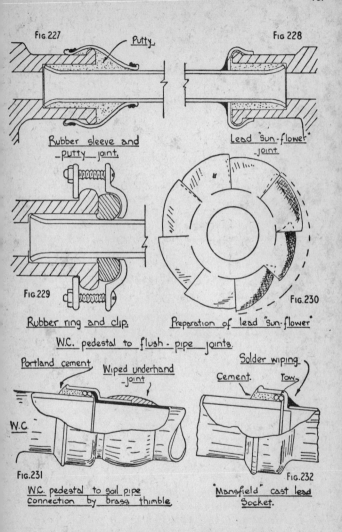

Fig. 227

Putty.

Fig. 228

Rubber sleeve and putty joint.

Lead "Sun-flower" joint.

Fig. 229

Rubber ring and clip.

Fig. 230

Preparation of lead "Sun-flower"

W.C. pedestal to flush-pipe joints.

Portland cement

Wiped underhand joint

Solder wiping

Cement.

Tow.

W.C.

Fig. 231

W.C. pedestal to soil pipe connection by brass thimble.

Fig. 232

"Mansfield" cast lead Socket.

a lavatory. The word comes from the Latin *lavateo*, meaning wash. In everyday speech it has come to mean also a wash-room. But the plumber retains the original nomenclature, and the wash-basin is the lavatory basin.

This appliance is generally of vitreous china, but for schools, works, etc., can be of glazed fireclay.

It is provided with two square holes, into which the hot and cold pillar taps are fitted (Fig. 111). In the bottom of the basin a countersunk hole is provided, into which the "combination plug and waste" must be fitted. High up on the back of basin, a weir overflow is connected by a passage in the thickness of the wall to a slot in the waste fitting (Fig. 112). At a convenient point a chain bolt is secured through a hole, and connected by means of a piece of chromium chain to the plug (Fig. 114).

Lavatory basins are of a number of types, some are fixed on a pedestal, some on brackets of various kinds, others are made to fit into the angle of a room, and yet others are assembled in ranges of four or six or more. Many existing basins are supported on cast-iron legs and stands, but this practice is rightly out of favour.

The basin should be fixed at a convenient height so that those who will use it will have maximum comfort. In general, a height of about 800 mm at the front is suitable.

Where brackets are to be used, they should be either the towel-rail type, screwed on to wooden pads or directly to wall plugs, or the cantilever kind built into the wall to a distance of 115 mm. Where the brackets are to be screwed on to wooden pads, two good long wooden plugs should be driven well into the brickwork joints—previously cleaned of mortar. The pads should be carefully nailed—using 75 mm (3 in) nails—so that the pad tops are level with each other, and form a ledge on which the back of the basin will rest. This precaution will prevent the whole weight of the basin being carried on the bracket screws. The pads should be 25 mm (1 in) thick, and the screws about 32 mm (1¼ in) and of substantial strength—say 12 or 14 gauge.

The waste fitting (Fig. 112) is made watertight by means of putty, all surfaces being painted. A bedding of putty is used under the flange of the fitting as softening and to make the plug an effective stopper. The watertight joint is made at the bottom of the basin. A ring of putty is pressed around the thread of the fitting, and a lead washer is inserted and followed up by the back-nut. The back-nut should be tightened firmly but carefully, with additional care that the overflow slot is in alignment with the over-flow opening in the basin. The lead washer should not exceed the outside measurement of the brass back-nut. Any surplus lead is useless and untidy because the watertight joint is made where the threaded fitting emerges from the basin. The tail-piece or union attached to the waste pipe should be made good by means of a leather washer or greased soft string. Once more care must be taken that the waste fitting does not turn in the process of tightening the union.

The pillar taps (Fig. 111) should be secured by means of a small quantity of plaster of Paris, a lead washer as softening, and the back-nut. The plaster when set prevents the square on the tap from turning in the somewhat larger square hole in the basin. The tap-fixing operation must be carried out quickly and methodically in the time before the plaster sets. The taps must not be moved for some time after fixing to allow the plaster to harden—say at least an hour.

New taps are supplied with fibre washers by which means the unions can be made good to the taps. Otherwise greased string can be used.

Fig. 113 shows a special fitting to be used with light-gauge copper tube. This replaces the usual "cap and lining" or union. Fig. 111 shows a lead pipe connected by joint wiping (Chapter XII). For handling the somewhat inaccessible back-nuts and union nuts *in situ*, a special "crowfoot" spanner is used.

Baths

Most baths are of vitreous-enamelled cast-iron; they have a square top and are fitted with panels on any exposed sides. There is a growing tendency to use pillar taps instead of the lately common globe-type taps, fixed on the vertical end of the bath. By using pillar taps, the bib or outlet can be raised above the flood rim of the bath, and any danger of water pollution in the service pipes—due to back siphonage—is made less possible. The taps are secured as in a lavatory basin. It is important that before any tap is fixed, the tap top should be removed and reassembled. It often happens that tap tops are abnormally tight, and a lavatory basin or any other fitment might be damaged in the struggle.

When the tap end of a bath is close to a wall, it is necessary to couple up the farthest tap first, then the overflow, and finally the near tap. Here again a cranked spanner is useful.

The bath waste should be fitted before the bath is lifted into position, and the waste pipe should be in place and complete with trap, so that only a horizontal nut remains to be tightened. The bath should be fixed before the plastering or tiling is done, so that a watertight joint is made with the wall.

The overflow is connected to the bath by means of a "male," i.e., outside threaded, elbow; the ornamental circular weir screwing on to it from the inside and made watertight by means of a hemp grummet or a washer of soft material. The overflow pipe can be branched into the waste pipe or may be taken through the wall separately. In this case a copper flap should be fitted to the end, to prevent an uncomfortable draught in the bath.

Sinks

Sink fashions are changing. We have passed from the old shallow "slop-stone" to the deeper "Belfast" sink and the heat-wasteful combination sink and drainer—all in glazed earthen-ware, to the much-to-be-desired stainless-steel sink. But the price is, and will remain, a deciding factor for most households.

The Belfast sink, measuring usually 600 × 400 × 250 mm (24 × 16 × 10 in) or thereabouts, is all-white or cane and white.

The combination sink measures about 750 × 400 × 250 mm (30 × 16 × 10 in) and is generally all-white finished.

Both these sinks are fitted with combination plug and waste (Fig. 112), and have a secret overflow with open weir inlet. The fixing procedure is as described in relation to lavatory basins.

The water-supply is from bib taps (Fig. 109). They should be fixed with the bib about 330 mm (13 in) from the inside of the bottom of the sink, so that a bucket can be filled without diffi-culty. This height should not be exceeded, or splashing will result. And loose anti-splash nozzles are to be recommended.

It is convenient to fix the taps 150 m (6 in) apart at the centres for tiling purposes. Bib taps should always be of the screwed-boss type (as shown) so that in case of replacement there is no necessity for damage to tiles or other wall finishes. The boss can be jointed to lead pipe by wiping, or if copper tube is used special elbows, tees, etc. are available.

As sinks are heavy and are often well loaded, they need good support. In the old days brick pillars were often used, but they are to be deprecated. The support should be such that there is no difficulty in cleaning the wall and floor. For this reason cantilever brackets should be used. Special brackets can be bought, but two pieces of angle- or tee-iron will do admirably; they should be well fastened in the brickwork preferably by building in. For neatness, the brackets should finish about 75 mm (3 in) from the front of the sink and be cut diagonally. For the same reason, they should be well away from the ends.

The back of the sink should be bedded in cement against the wall, so that the nuisance of water running down the wall is avoided.

Stainless-steel sinks are fitted with pillar taps or mixer units and require the same treatment as lavatory basins. They have a great advantage over glazed earthenware, in that less heat is taken from the water by the sink. This factor removes the need for using an enamel bowl for washing dishes, a practice so common with large earthenware sinks.

CHAPTER XIX

SOIL AND WASTE DISPOSAL

Traps for Sanitary Fittings

ALL sanitary fittings are fitted with traps to prevent the air contained in drains and sewers from passing into a house by way of the soil or waste pipes. Unpleasant smells and disease germs are the dangers.

Suitable traps for water-closets are shown in Fig. 224; they form an integral part of the fitting.

Traps for lavatory, bath, and sink are illustrated in Figs. 233–237. The S and P traps will accommodate most waste-pipe requirements.

The running-trap (Fig. 235) is useful where it is advantageous to put a trap on a horizontal waste pipe.

Where an "in-line" trap is needed, the bag-trap or the neater bottle-trap can be used. The latter has the advantage that it can easily be taken apart and cleaned out. All traps should be provided with a means of access.

Waste Pipes

Waste pipes from lavatory basins, baths, and sinks may be executed in lead (traditional), copper, and cast-iron. Individual waste pipes should take the shortest path, sharp bends should be avoided, and they should be of such size that they will generally run fairly full, and thereby be self-cleansing.

Groups of fittings may be dealt with by such arrangements as shown in Figs. 238–240, where the fittings may be trapped singly or as a group.

Lead waste pipes are jointed by means of wiped joints. Underhand joints are recommended, but in practice flange joints are often used. The bending of solid-drawn lead pipe is dealt with in Chapter XIV. Whenever lead pipe is used, it should be properly supported to avoid sagging. Suitable methods are shown in Figs. 122, 130, and 131.

Copper waste pipes, being more rigid than lead, require less support. Copper tube can be bent to requirement as described in Chapter XV. For jointing, compression fittings, capillary fit-

Traps for sanitary fittings.

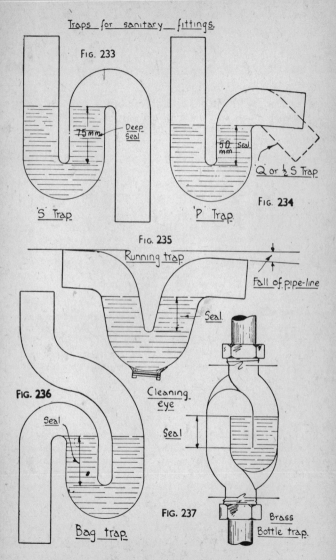

FIG. 233

75mm — Deep Seal

'S' Trap

FIG. 234

50 mm Seal

Q or ½ S Trap

'P' Trap.

FIG. 235

Running trap

Fall of pipe-line

Seal

Cleaning eye

FIG. 236

Seal

Bag trap.

FIG. 237

Seal

Brass Bottle trap.

tings, and weldings can be used. Fig. 238 illustrates the use of compression fittings on a range of three lavatory basins.

Cast-iron pipes are sometimes used. They should be of heavy-quality, treated inside and out with Dr. Angus Smith's Bituminous Solution. The joints should be caulked with tow and lead-wool or molten lead. Suitable bends and junctions can be acquired. Where a lead pipe is to be connected to cast-iron, the joint shown in Fig. 245 should be used. It will be seen that the lead pipe passes through a brass sleeve, this sleeve enables the joint to be caulked.

Waste pipes should be laid to proper falls, and access should be provided in order that each length of pipe can be rodded (Fig. 251).

Soil Pipes

Soil pipes are those which convey the contents of water-closets and urinals to the drains. The vent-stack which is normally carried above the roof is also included in any discussion of soil pipes.

In general, a soil-disposal system is a combination of lead and cast-iron pipe work. Water-closet outlets usually discharge through the outside wall by means of a solid-drawn lead bend into a cast-iron junction in a cast-iron ventilation shaft. The connection of the pedestal to the lead pipe has already been discussed (Figs. 231 and 232). The lead-to-iron joint is shown in Fig. 245. For the caulking, lead-wool may be used instead of molten metal.

Every soil pipe must terminate in a rest-bend on the drain; the cemented joint can be seen in Fig. 246. The soil pipe should be carefully centred by means of the tow caulking, otherwise a faulty joint may result.

All cast-iron soil-pipe joints must be caulked with lead to ensure reliable joints. Like rain-water pipes, cast-iron soil pipes are provided with ears for nailing, or in special cases with holder-bat fittings. In this case the holder-bat is built into the wall and the pipe attached to it by means of a dove-tail or a bolted collar. All soil pipes should be coated with Dr. Angus Smith's Solution. This is a protective process in which the pipe when hot has been dipped in a special solution of pitch, resin, etc. The solution is very durable and to be preferred to any form of paint.

Lead-pipe bends can be purchased ready made, but they can be

Lavatory basin waste connections

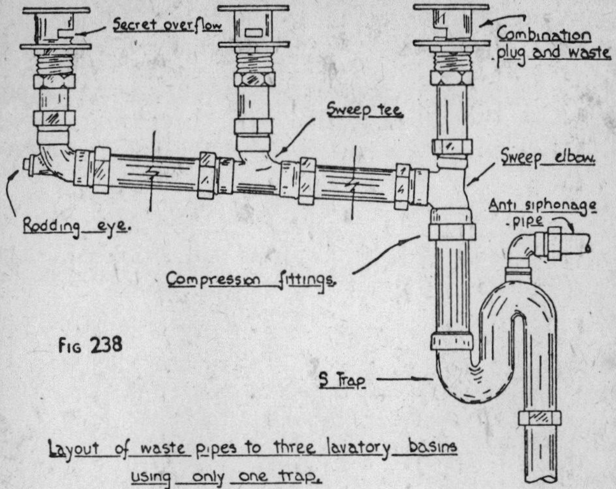

- Secret overflow
- Combination plug and waste
- Sweep tee
- Sweep elbow
- Anti siphonage pipe
- Rodding eye.
- Compression fittings.
- S Trap

Fig 238

Layout of waste pipes to three lavatory basins using only one trap.

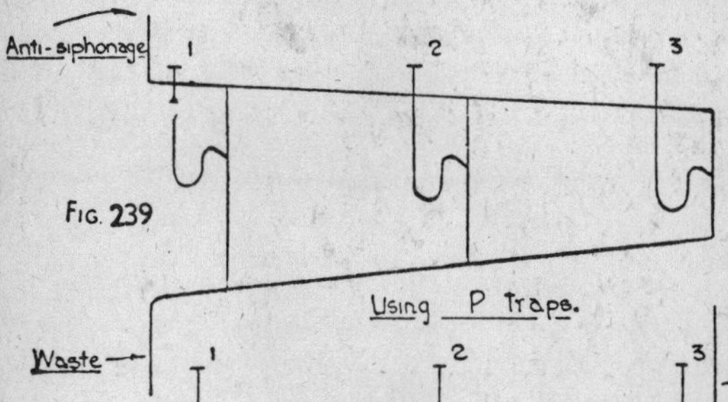

- Anti-siphonage   1   2   3
- Fig. 239

Using P Traps.

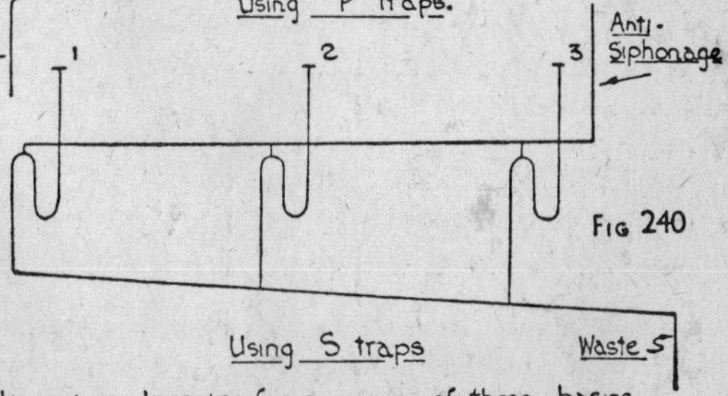

- Waste   1   2   3   Anti-Siphonage

Fig 240

Using S traps   Waste

Alternative layouts for a range of three basins.

Soil and waste disposal.

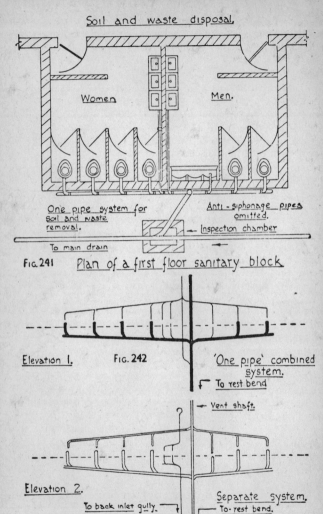

Women

Men.

One pipe system for Soil and waste removal.

Anti-siphonage pipes omitted.

← Inspection chamber

To main drain

Fig. 241   Plan of a first floor sanitary block

Elevation 1.   Fig. 242

'One pipe' combined system.

⌐ To rest bend

← Vent shaft.

Elevation 2.

Separate system.

To back inlet gully →

To rest bend.

Fig. 243

made more cheaply by skilled craftsmen and they lack the firmness of a hand-made bend. Lead-pipe bending is dealt with fully in Chapter XIV.

Anti-siphonage

Where the discharge from one w.c. passes the branch taking another w.c., the pressure may be so reduced in the branch that atmospheric pressure on the surface of the water in the basin may force the contents down the pipe. This occurrence is common where two w.c.s are on the same pipe or where w.c.s are situated above each other and discharge into the same stack.

This indicates the need for precautionary measures, and wherever this position exists anti-siphonage pipes are introduced. Fig. 241 is the plan of a sanitary block on the first floor of a building. The elevation 2 (Fig. 243), shows how six w.c.s and a urinal discharge into main soil pipes which lead into a vent shaft and soil pipe. The waste pipes from the two lavatory-basin ranges can also be seen conducted separately to the drains. This is known as a separate system of soil and waste disposal, soil and waste being kept entirely separate.

The anti-siphonage provision is also shown in Fig. 243, a branch pipe being taken from each fitment to lead back into the vent shaft. By this means there is an air-inlet near the outlet of each pedestal trap, so that equal pressure on each side of the water-seal in the traps is ensured.

Some w.c. pedestals are provided with a socket in a position close to but not at the crown of the trap, into which the anti-siphonage pipe can be fitted. This is useful where the anti-siphonage pipes are carried on the inside of a wall, and in cases where local by-laws require connection within 75 mm and 460 mm (3 in and 18 in) of the crown of the trap (Fig. 248).

Some local by-laws, on the other hand, are sufficiently elastic to allow the connection to be made where the soil pipe emerges from the wall on the outside (Fig. 247). In this case the anti-siphonage is just as efficient, and there is the advantage that only one hole is cut in the wall.

The Model Building By-laws, on which many local regulations are based, require that the connection shall be made not less than 76 mm (3 in) and not more than 300 mm (12 in) from the crown of the trap of the w.c. pedestal. In this case the connection must be made inside the building or within the thickness of the wall.

Soil pipe joints.

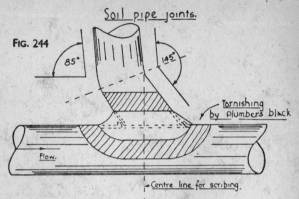

Fig. 244

85°

145°

Tarnishing
by Plumbers black

Flow.

Centre line for scribing.

Angle of bend and setting out of branch joint on lead soil pipe

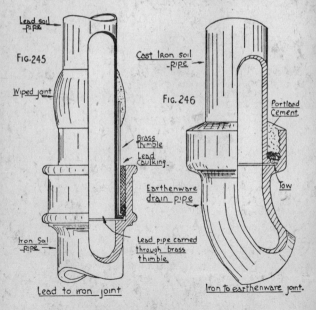

Lead soil pipe

Fig. 245

Wiped joint

Cast Iron soil pipe

Fig. 246

Portland Cement.

Brass Thimble

Lead Caulking.

Earthenware drain pipe

Tow

Iron Soil pipe

Lead pipe carried through brass thimble

Lead to iron joint

Iron to earthenware joint.

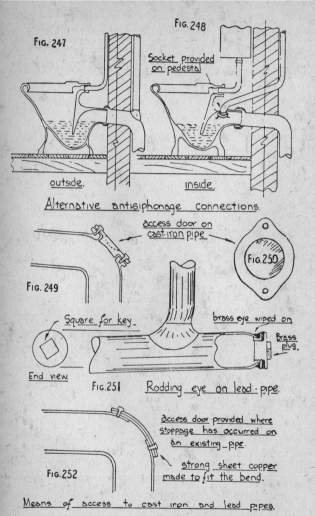

FIG. 247

FIG. 248

Socket provided on pedestal

outside.

inside.

Alternative antisiphonage connections.

access door on cast-iron pipe

FIG. 249

FIG. 250

Square for key

brass eye wiped on

Brass plug.

End view.

FIG. 251   Rodding eye on lead-pipe.

access door provided where stoppage has occurred on an existing pipe.

strong sheet copper made to fit the bend.

FIG. 252

Means of access to cast iron and lead pipes.

Just as soil pipes must have a regular fall towards the drains, the anti-siphonage pipes must rise until they reach the vent pipe.

On soil and anti-siphonage pipes all branches should be swept in the direction of the flow (Fig. 244). 100 mm (4 in) soil pipes should fall at the rate of 1 in 40, 150 mm (6 in) pipes àt 1 in 60. For the sake of appearance, the anti-siphonage pipes should have a comparable slope.

The Combined or "One-pipe System"

In the "one-pipe system" all sanitary fittings discharge into the same soil pipes with an economy of pipe work, and a simplification of layout. A few precautions need to be taken.

All lavatory basins, baths, and sinks must be fitted with deep-seal traps or with anti-siphonage pipes.

The anti-siphonage pipe can be returned into the vent shaft at a point not less than 1 m (3 ft) above the highest soil pipe junction.

The main anti-siphonage pipe must, in certain circumstances, be carried down and connected into the main soil stack below the lowest inlet branch. This precaution is necessary where fittings are situated on two or more floors and discharge into a common stack. The arrangement is shown in Fig. 242 for convenience. Its purpose is to counteract compression set up when a body of air is caught between two columns of water. If this should happen at a branch, the seal may be blown by the pressure produced.

Means of access should be provided so that all pipes can be rodded. Suitable methods are shown in Figs. 249–252.

INDEX

TEACH YOURSELF BOOKS

HOME HEATING

B. J. King & J. E. Beer

Intended as a guide to any householder contemplating buying a heating system for his home, this book discusses the advantages and limitations of all the different methods of heating a house.

Basic principles and terminology are first explained, and then the different types of heat generators – gas, oil, electric and solid fuel – and the systems with which they are used. Hot water systems, radiators, convectors, warm air units, skirting, ceiling and underfloor heating are all covered, and there are also chapters on controls and insulation, and on buying and maintaining the system. Illustrated throughout, the text includes useful appendices listing the names and addresses of equipment manufacturers.

With the help of this book, home-owners will be able to choose the heating system most suited to their individual needs.

UNITED KINGDOM   40p
AUSTRALIA $1.25*
NEW ZEALAND $1.10
CANADA $1.50
*recommended but not obligatory

ISBN 0 340 16428 X

TEACH YOURSELF BOOKS

HOUSE REPAIRS

T. Wilkins

Anyone owning a house needs to protect his investment.
This book shows you how to maintain and keep in good
repair your property, without having to call in expensive
outside help.

A do-it-yourself manual, covering all the jobs that arise
in and around the house, *House Repairs* is full of practical
advice and information for the handyman. Working
illustrations are used throughout and the reader should
have no difficulty in following the instructions given on
repairs, to roofs, floors, windows, etc.

An invaluable guide for any householder by the Editor of
Do-it-Yourself magazine.

UNITED KINGDOM	40p
AUSTRALIA	$1.40*
NEW ZEALAND	$1.40
CANADA	$1.50

ISBN 0 340 05624 X   *recommended but not obligatory